New Concepts in Polymer Science

Biocide Guanidine Containing Polymers:
Synthesis, Structure and Properties

New Concepts in Polymer Science

Previous titles in this book series:

Active Sites of Polymerization.
Multiplicity: Stereospecific and Kinetic Heterogeneity
Y.B. Monakov, N.N. Sigaeva and N. Urazbaev

Stationary and Non-Stationary Kinetics of the Photoinitiated Polymerization
Yu.G. Medvedevskikh, A.R. Kytsya, L.I. Bazylyak, A.A. Turovsky and G.E. Zaikov

Thermal Stability of Engineering Heterochain Thermoresistant Polymers
E.V. Kalugina, K.Z. Gumargalieva and G.E. Zaikov

The Concept of Micellar-Sponge Nanophases in Chemical Physics of Polymers
Yu.A. Mikheev, G.E. Zaikov

Structure of the Polymer Amorphous State
G.V. Kozlov and G.E. Zaikov

The Statistical Nature of Strength and Lifetime in Polymer Films and Fibers
B. Tsoi, E.M. Kartashov and V.V. Shevelev

Cyclolinear Organosilicon Copolymers: Synthesis, Properties, Application
O.V. Mukbaniani and G.E. Zaikov

Modern Polymer Flame Retardancy
S.M. Lomakin and G.E. Zaikov

Polymeric Biomaterials.
Part I. Polymer Implants
M.I. Shtilman

Biodegradation and Durability of Materials under the Effect of Microorganisms
S.A. Semenov, K.Z Gumargalieva and G.E. Zaikov

Fire Resistant and Thermally Stable Materials Derived from Chlorinated Polyethylene
A.A. Donskoi, M.A. Shashkina and G.E. Zaikov

Polymers and Polymeric Materials for Fiber and Gradient Optics
N. Lekishvili, L. Nadareishvili, G. Zaikov and L. Khananashvili

Polymers Derived from Isobutylene. Synthesis, Properties, Application
Yu.A. Sangalov, K.S. Minsker and G.E. Zaikov

Ecological Aspects of Polymer Flame Retardancy
S.M. Lomakin and G.E. Zaikov

Molecular Dynamics of Additives in Polymers
A.L. Kovarski

Structure and Properties of Conducting Polymer Composites
V.E. Gul'

Interaction of Polymers with Bioactive and Corrosive Media
A.L. Iordanskii, T.E. Rudakova and G.E. Zaikov

Immobilization on Polymers
M.I. Shtilman

Radiation Chemistry of Polymers
V.S. Ivanov

Polymeric Composites
R.B. Seymour

Reactive Oligomers
S.G. Entelis, V.V. Evreinov and A.I. Kuzaev

Diffusion of Electrolytes in Polymers
G.E. Zaikov, A.L. Iordanskii and V.S. Markin

Chemical Physics of Polymer Degradation and Stabilization
N.M Emanuel and A.L. Buchachenko

New Concepts in Polymer Science

Biocide Guanidine Containing Polymers:

Synthesis, Structure and Properties

Nikolai A. Sivov

CRC Press
Taylor & Francis Group
Boca Raton London New York

CRC Press is an imprint of the
Taylor & Francis Group, an **informa** business

First published 2006 by Koninklijke Brill

Published 2022 by CRC Press
Taylor & Francis Group
6000 Broken Sound Parkway NW, Suite
300 Boca Raton, FL 33487-2742

ISBN-13: 978-90-6764-447-1 (hbk)

Visit the Taylor & Francis Web site at
http://www.taylorandfrancis.com

and the CRC Press Web site at
http://www.crcpress.com

A C.I.P. record for this book is available from the Library of Congress

CONTENTS

Preface

Synthetic polyelectrolytes play a very important role in science, technique and medicine at present. At the same time constant broadening of application spheres of polymers of this class is observed. As a consequence we observe the growth of requirements for creation of novel polyelectrolytes – polymers and copolymers of prescribed chemical and stereo-chemical structure and molecular mass. This fact stimulates investigations of problems of synthesis and formation mechanism of this class of polymer compounds. It is also obvious that the most simple and convenient methods of polyelectrolytes reception are the reactions of ionized monomers radical polymerization and copolymerization.

The presented book is devoted to a number of problems connected with polyelectrolytes chemistry including biocyde guanidine containing polymers. The main attention in this book is paid to the results of studies carried out by the author of this book and under his direction in the Laboratory of Polyelectrolytes and Surface-active Polymers Chemistry of A.B. Topchiev's Institute of Petrochemical Synthesis (IPS) of Russian Academy of Sciences (RAS). In particular, the main topics are: synthesis of novel guanidine containing diallyl and vinyl monomers, investigation of their physical-chemical properties and synthesis of novel water-soluble (co)polymers possessing wide range of physical-chemical characteristics and biocyde properties by radical polymerization on the base of obtained results.

The principles of synthesis and the number of methodic recommendations on reception of guanidine containing monomers are set in details in Chapter 1 and in the Appendix. The main regularities of radical polymerization of synthesized monomers are presented in Chapter 2 (namely the original studies are described in sections 2.1.2 and 2.2.2). The results of successful production of copolymers of various structures on the base of synthesized guanidine containing monomers of diallyl and vinyl type are considered in Chapter 3 sections 3.2 and 3.3. Chapter 4 is devoted to extensive physical-chemical study of synthesized for the first time monomer and polymer products where the correlation between physical-chemical properties and structure of given compounds with their behavior is tracked. Separate Chapter 5 is devoted to biocide properties of polyelectrolytes of some known and novel guanidine containing compounds which were synthesized by the author of given book or under his direction for the first time.

Author considered necessary to make in the book the list of abbreviations which would make easier understanding of presented work. In the Appendix at the end of the book readers will find useful for synthesis details and additional data concerning various methods used in the work. While analyzing theoretical and practical aspects of given processes author used various literature sources, corresponding references were presented at the end of each Chapter.

Author hopes that this book will be helpful for researchers engaged in polyelectrolytes chemistry, process of their formation, investigation of their various physical-chemical and biological properties.

Author expresses his thanks to those colleagues who made a great contribution to preparation of this book:

to Martynenko A.I., Kabanova E.Yu. Popova N.I. – research collaborators of author's Laboratory of Polyelectrolytes Chemistry of A.B. Topchiev's Institute of Petrochemical Synthesis of RAS who carried out a great number of experiments and helped in results interpretation;

to Bondarenko G.N. (research officer of A.B. Topchiev's Institute of Petrochemical Synthesis RAS) assisted in carrying out of IR-studies and analysis of obtained results;

to Philatova M.P. who measured numerous NMR spectra;

to Krut'ko E.B. who established elemental structures of monomer and polymer compounds;

to author's teachers Plate N.A., Kabanov V.A., Topchiev D.A. who inspired author to write this book by their example and knowledge;

to my family, to my wife Ludmila and sons Alexander and Dmitry for their help in preparing of this book.

I want to tender particular thanks to Zaikov G.E. for important advices under preparing of this book.

Abbrevations

AA	Acrylic acid
AAm	Acrylamide
ADMA	Allyldimethylamine
AG or GA	Acrylate guanidine or Guanidine acrylate
AIBN	Dinitrilazobis isobutyric acid
APS	Ammonium persulfate
CA	Cyanamide
DAA	Diallylamine
DADAAH	Diallyldialkylammonium halogenide
DADMA	Diallyldimethylammonium
DADMAC	N,N-diallyl-N,N-dimethylammonium chloride
DAG	Diallylguanidine
DAGA	Diallylguanidine acetate
DAGTFA	Diallylguanidine trifluoroacetate
DAMA	Diallylmethylamine
DAMATFA	Diallylmethylamine trifluoroacetate
DCT	Degradational chain transfer
DEAEMA	N,N-diethylaminoethyl methacrylate
DMFA	Dimethyl formamide
DMSO	Dimethylsulfoxide
DMSO-d_6	Deuterated dimethylsulfoxide
DNA	Deoxyribonucleic acid
DSC	Differential scanning calorimetry
DTA	Differential thermal analysis
DTG	Differential thermal gravimetry
EDA	Ethylene diamine
EG	Ethylene glycol
ETED	Endothermic effect of decomposition
FAA	Fluoroacrylic acid
GHC	Guanidine hydrochloride
GP	Triguanidine phosphate
IBA	Isobutylamine
IPEC	Interpolyelectrolyte complex
IR	Infra-red
MAG	Methacrylate guanidine
MAA	Methacrylic acid
MSC	Minimum suppression concentration
NMR	Nuclear magnetic resonance
PAA	Polyacrylic acid
PAG	Polyacrylate guanidine
PAS	Physiologically active substances
PDADMAC	Poly-N,N-diallyl-N,N-dimethylammonium chloride
PE	Polyelectrolyte
PHMG HC	Polyhexamethylene guanidine hydrochloride
PMAA	Polymethacrylic acid
PMAG	Polymethacrylate guanidine
PVP	Poly-N-vinyl pyrrolidone
TEA	Triethylamine
TFAA	Trifluoroacetic acid

UV	Ultra-violet
2VP	Vinyl pyridine
VP, N-VP	N-vinyl pyrrolidone

Introduction

Creation of novel biocide polymers represents important direction in present-day chemistry of high-molecular compounds. This problem became especially relevant recently when wide spreading of stable to various bactericidal substances strains and possibility of their epidemic spreading was the serious problem for the formation of effective anti-bacterial therapy.

That is why the search of means providing blocking of several factors of stability of pathogenic microorganisms becomes necessary. Application of polymers able to reveal combined effect on bacterial cell which are more effective and less dangerous for person in comparison with low-molecular biocide substances traditionally used for protection of microorganisms seems to be perspective for the solution of this problem.

The important aspect of this problem is the developing of biocide polymers and scientific bases of their synthesis, and consequently investigation of quantitative physical-chemical regularities of polymers formation processes, finding of possibilities for sensitive regulation under synthesis of polyelectrolytes of polymers' structure, composition, distribution of chemical units in macromolecules, molecular masses and properties, including biocide ones.

Quaternary ammonium salts of diallyl type and acryl acids' salts are perspective chemical structures for reception of novel guanidine-containing biocide polymers. The choice of namely these compounds as subjects of inquiry was caused by the following.

Firstly, it is well known that acids of acryl type containing chemically active functional groups represent perspective type of monomers since polymers and copolymers on their base may keep activity potential being suitable carrier of various substances including biologically active ones. Moreover, they posses the wide range of practically important properties and are used in pulp and paper, varnish-and-pain, textile industries as emulsifying agents, for dressing of synthetic fibers, for preparing of latexes, glue compositions, etc.. Homo- and copolymers on the base of well-known cationogeneous monomer N,N-diallyl-N,N-dimethylammonium chloride (DADMAC) are traditionally and widely used in various branches of industry, including oil-producing and pulp and paper, under waste treatment, in ceramic and silicate production. Moreover, recently they are used in electronics, production of contact lenses, in agriculture for improvement of soils structure and other aims.

Secondly, the presence in elementary unit of these polymers of guanidine group should impart them high biocide activity. It is well known that compounds containing guanidine group in their structure possess wide spectrum of bactericide action and are used as medicines including antibiotics. As investigations of last years showed such polymers might reveal combined influence on bacterial cell being at that more effective and less dangerous in comparison with low-molecular biocide substances traditionally used for microorganisms protection. Especially valuable property of guanidine containing compounds is that in the alive organism there are ferment systems able to destruct these compounds and prevent their cumulation. However, one should remember that for every novel guanidine-containing compound or a number of compounds close in structure the ability of effective work of such ferment systems requires additional verification.

Thirdly, guanidine containing polymers except polyhexamethylenguanidine (PHMG) and its derivatives are practically not studied in literature. That is why a lot of scientific and technological aspects of guanidine containing polyelectrolytes synthesis are solved insufficiently that hinders creation of novel biocide preparations on their base. (We should note that developing of laboratory methods and industrial ways of PHMG synthesis is made in the laboratory of Polyelectrolytes Chemistry and Surfactant Polymers of A.V. Topchiev's Institute of petrochemical synthesis of Russian Academy of Sciences which is now directed by the author of this book.)

And finally, under synthesis of guanidine containing polymers there are definite technological difficulties. Known guanidine-containing polymers are prepared by polycondensation as a consequence of which not high molecular masses are characteristic for them (10-12 thousands). Synthesis of guanidine containing polymers by radical polymerization allows obtaining of polymers with wide variety of molecular masses values (from 20 thousands up to the million and higher) and regulating of such polymers' structures and properties in wide range.

It is also important that the presence in given polymers of functional groups able to various modifications significantly increases possibilities of macromolecular design.

On the base of said above we may conclude that synthesis of novel guanidine containing polyelectrolytes on the base of monomers of acrylic type and dialkyldiallylammonium derivatives with the use of radical polymerization and investigation of scientific bases of these processes, and also the developing of methods of novel guanidine containing biocide polyelectrolytes reception are very relevant problems.

Chapter 1. Synthesis of guanidine containing monomers

The selection of monomers for creation of novel polymer products was conditioned by various reasons, first of all by the necessity of presence of guanidine and various ionogenic groups. As is well known the compounds containing guanidine groups reveal noticeable biocide activity. Introduction of such groups into polymer chains leads to strengthening of biocide properties, that is why creation of novel polymers containing guanidine groups, investigation their physical-chemical properties and characteristics, revealing of possibility of preparation of polymer products possessing biocide properties on their base are very important today.

Some derivatives of diallyl (N, N–diallylamine) and vinyl (acrylic acids) series were chosen for solution of this problem as initial objects of novel monomers synthesis.

1.1. Synthesis of diallyl guanidine containing monomers

Synthesis of high-molecular products by radical polymerization on the base of monomers of allyl series is connected with fundamental difficulties due to proceeding of degradation chain transfer (DCT) on monomer in polymerization course. Additional "modification" of diallyl monomers, namely creation of stable monomers' protonated forms in polymerization system (i.e. formation of monomeric salt with positive charge on nitrogen atom) is necessary for inhibiting of DCT. The presence of positive charge on nitrogen atom is in particular the key factor for solution of considering problem.

The systems in which monomers of diallyl series containing guanidine groups exists in protonated form were found and presented in given work. Guanidine-containing monomeric salts on the base of diallylguanidine (DAG) and organic acids (acetic and trifluoroacetic ones) were synthesized with this aim (Scheme 1.1).

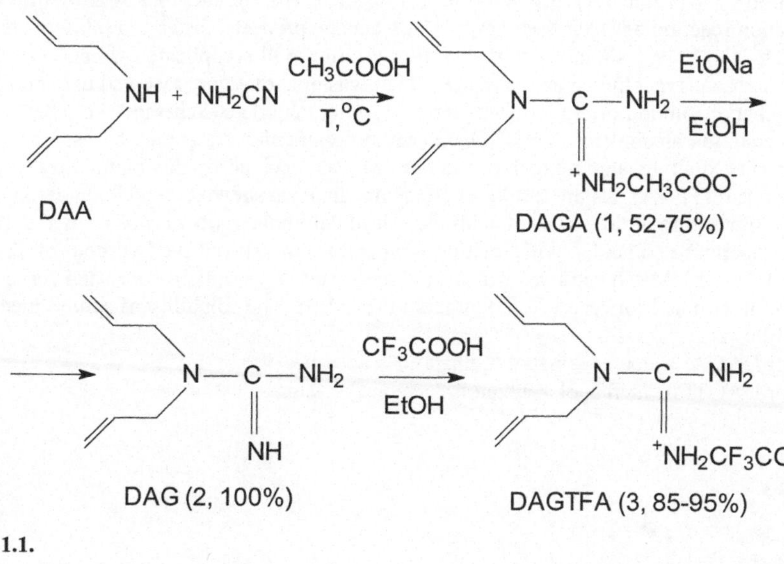

Scheme 1.1.

The simplicity and convenience of technique suggested by us consist in the following. The mixture of diallylamine (DAA) and DAA acetate preliminary prepared under cooling from diallylamine (DAA) and acetic acid was mixed with solution of cyanamide (CA) in acetonitrile. The amine was got in some excess, since at the first stage the nucleophilic addition of DAA by nitrile group of CA occurred (Scheme 1.2).

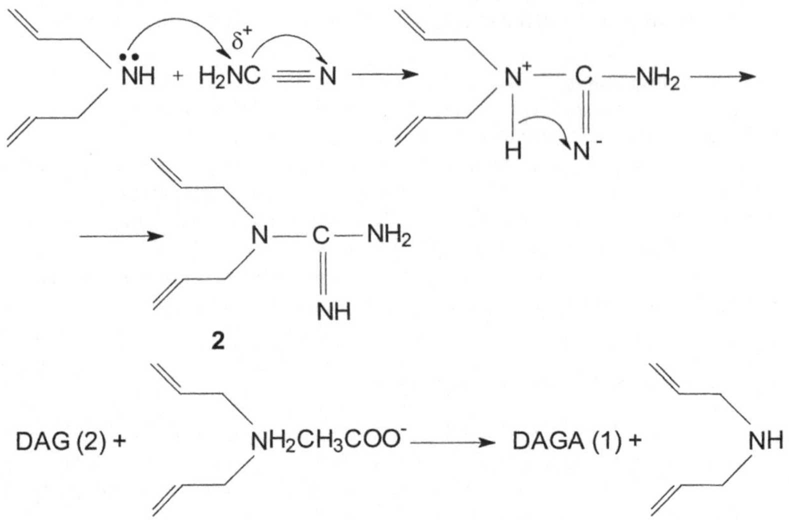

Scheme 1.2.

The resulted terminal mixture was gradually dropped into boiling acetonitrile. Received DAG (2) was unstable at high temperatures compound but possessing higher basicity than DAA it reacted with DAA acetate and formed stable salt form of diallylguanidineacetate (DAGA) (1) releasing the DAA for further conversions. As investigations showed the increase of volume of boiling acetonotrile (from 25 up to 250 ml per 1 mole of CA) increased the yield of desired product DAGA (1) from 52% up to 75% (Table 1.1). Probably, it is connected with the fact that in more diluted solutions protonation reaction of DAG base (2) by DAA acetate proceeds faster and totally with formation of DAGA salt (1). We should note that further dilution will complicate extraction of product. Because of the reasons mentioned above the DAG (2) was quantitatively received under addition of DAGA into ethanol solution of preliminary prepared sodium ethylate (Scheme 1.1). After removing of sodium acetate the alcohol solution of DAG (without extraction of it) may be used for preparation of other salts; diallylguanidine trifluoroacetate (3) was received by this method by addition of trifluoroacetic acid (TFAA). As the results of elemental analysis showed, in wide diapason of ratios DAG : TFAA (up to three-fold excess of acid) the salt of equimole structure was formed at that (and its melting temperature coincided with melting temperature of salt received without of TFAA excess) (Table 1.2). ^{1}H NMR spectra were measured and solubility limits in water and some organic solvents were determined for received monomers (Table 1.3). The solubility of synthesized salts is changed as follows:

1. for DAGA: water > methanol > ethanol >> acetone;
2. for DAGTFA: methanol > ethanol > acetone > water.

Table 1.1. Synthesis of diallylguanidineacetate (DAGA)

No.	DAA[1], ml (g, mole)	AC[2], ml (g, mole)	CA[3], g (mole)	Acetonitrile[4], ml	Acetonitrile[5], ml	t[6], min	Yield[7], %
1	127 (100,5, 1,034)	57 (59,9, 0,986)	41,4 (0,986)	50	25	100 + 110	52
2	169 (133,4, 1,375)	75 (78,6, 1,310)	55,8 (1,310)	65	170	120 + 90	67
3	134 (105,9, 1,070)	59 (62,0, 1,020)	43,9 (1,020)	50	250	100 + 90	75
4	150 (118,7, 1,224)	67 (69,9, 1,186)	49,0 (1,186)	60	250	105 + 105	73
5	157,5 (124,4, 1,280)	70 (73,5, 1,220)	51,3 (1,220)	60	250	105 + 105	75

Note: [1] – diallylamine; [2] – glacial acetic acid; [3] – cyanamide; [4] – the volume of acetonitrile in which CA was preliminary dissolved; [5] – the volume of boiling acetonitrile into which the preliminary prepared mixture was dropped; [6] – the time of reaction: the first figure – duration of dropping, the second one – duration of additional boiling; [7] – the yield counting on CA.

Table 1.2. Solubility of novel monomeric salts of diallylguanidine

Solvent	H_2O		CH_3OH		C_2H_5OH		Acetone		EG		DMSO	
T, °C	20	60	20	60	20	60	20	60	20	60	20	60
Solubility of monomeric salts, mass %[1]												
DAGA	36	50	28	35	18	20	m	m	–	–	m	m
DAGTFA	7,7	20	31,3	53	18	35	11	25	25	35	–	–

Note: [1] m – marginally soluble, "–" – there are no data

Table 1.3. Elemental structure and melting temperatures of novel monomeric salts of diallylguanidine

Monomers	MM	T_{boil}, °C	Content of element, % Calculated / Practically observed			
			C	H	N	F
DAGA	199,253	211–212	54,25 / 53,84	8,60 / 8,80	21,09 / 22,24	
DAGTFA	253,224	157–158	42,69 / a)42,40 b)42,70	5,57 / a)5,55 b)5,60	28,95 / a)29,10 b)29,40	22,51 / a)21,82 b)22,30

Note: a) synthesis was carried out at equimolar of DAG-to-TFAA ratio; b) the three-fold excess of TFAA was used.

1.2. Synthesis of vinyl guanidine containing monomers

Guanidinemethacrylate (MAG, 4) and guanidineacrylate (GA, 5) were synthesized by us for the first time for solution of problem put by in this work and for widening of row of possible biocide objects [2]. Synthesis of these monomeric salts was carried out in two stages (Scheme 1.3):

1. synthesis of guanidine and its salt (in this case the guanidinehydrochloride was used, but usage of other guanidine salts was also possible; reaction 1);
2. synthesis of monomeric salts of guanidine and acrylic acids (reaction 2).

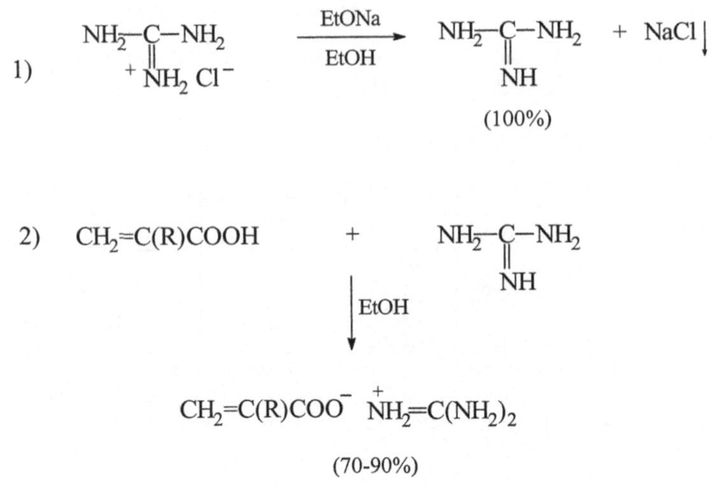

MAG: R = CH_3; **AG**: R = H

Scheme 1.3.

Synthesis of given monomeric salts was also carried out in isopropanol and butanol with corresponding sodium alcoholates for adjustment of GA and MAG synthesis conditions and it was shown that maximum yield (up to 90% in the case of MAG) was observed in isopropanol.

Synthesized monomeric salts of acrylic acids were studied by spectrum methods (^{1}H NMR and IR), and also by thermophysical methods. The results of these investigations will be described in details in Chapter 4. The structure of resulted monomeric salts was also confirmed by the data of elemental analysis (Table 1.4), solubility limits were also determined for them in water and number of organic solvents (Table 1.5).

Table 1.4. Elemental structures and melting temperatures of novel GA and MAG monomeric salts.

Monomers	MM	T_{boil}, °C	Content of element, % Calculated / Practically observed		
			C	H	N
GA	131,134	175–176	36,66 / 36,55	6,92 / 6,90	32,04 / 31,85
MAG	145,16	161–163	41,37 / 40,35	7,63 / 7,60	28,95 / 29,40

Table 1.5. Solubility of novel GA and MAG monomeric salts.

Solvent	H_2O		CH_3OH		C_2H_5OH		C_3H_8OH		Acetone	
T, °C	20	60	20	60	20	60	20	60	20	60
Solubility of monomeric salts, mass %[1]										
GA	57	87	40	44	33	40	22	32	ins	ins
MAG	62	80	55	62	48	55	38	41	ins	ins

Note: [1] ins – insoluble.

References

1. D.A. Topchiev, N.A. Sivov, E.J. Goethals, *Russian Chem. Bull.*, **43**, No.11, 1864 (1994).

2. N.A. Sivov, A.I. Martynenko, E.Yu. Kabanova, S.Yu. Khashirova, A.M. Esmurziev, *Neftekhimiya*, 1, 47 (2004).

Chapter 2. Radical polymerization of guanidine containing monomers

2.1. Radical polymerization of diallyl monomers

2.1.1. Radical polymerization of nitrogen containing diallyl monomers

Synthetic polyelectrolytes attract more and more attention of investigators recently as from theoretical, so from practical point of view. They play the important role in science, technique, medicine. At that we observe constant expansion of application fields of polymers of this class. This implies the growth of requirements to production of novel polyelectrolytes (PE), polymers and copolymers of given chemical and stereochemical structure and molecular mass that stimulates studies concerning structure, synthesis and mechanism of formation of various representatives of this class of polymer compounds.

It is obvious that in technological aspect the most simple and useful ways of PE production are reactions of radical homo- and copolymerization of ionized monomers [1, 2].

Systematic quantitative investigations of radical polymerization reaction of wide range of ionogenic monomers showed that these processes were characterized by specific kinetic particularities and mechanism of proceeding in comparison with non-ionogenic monomers that first of all was connected with the ability of initial monomers to form charged particles [2, 3]. Found differences and effects were characterized by the wide spectrum of additional influences the consideration and purposeful use of which opened practically analogously broad possibilities of reactions control as in the case of ion polymerization.

Undoubtedly, the important place in the raw of cationic PE is taken by monomer nitrogen-containing quaternary salts of diallyl raw which polymerization products may be used as coagulants and flocculating agents, soils' structurates, for production of ion-exchange sorbents, biocide compounds and in a lot of other cases.

We should note that problem of involving of ionogenic monomers of allyl type into radical polymerization is connected with well known limits caused by the reaction of degradational chain transfer (DCT) to monomer which is characteristic for all studied compounds of given raw (for example Scheme 2.1: reaction (1) – chain propagation; reaction (2) – degradational chain transfer on monomer).

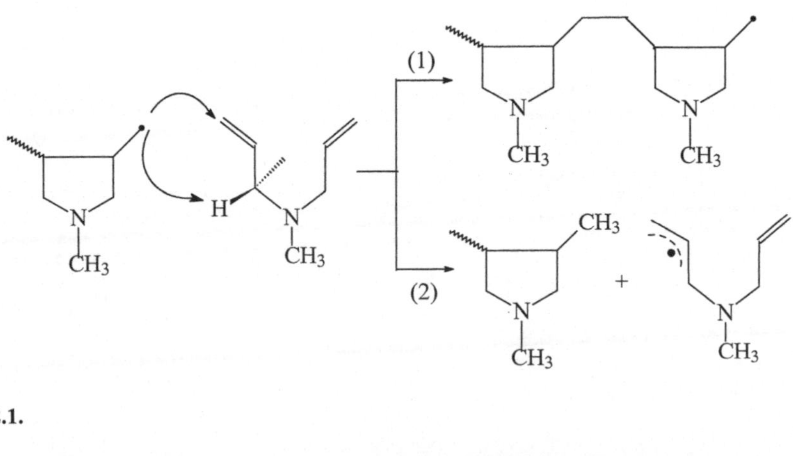

Scheme 2.1.

Overwhelming of these limits is partly possible with the help of complex formation agents [4] but at that the preparational and technological difficulties are naturally unavoidable.

The investigation of particularities of radical polymerization of allyl monomers of derivatives of quaternary ammonium salts [5] showed that at chosen conditions (radical initiation, medium protonation including H_2O) monoallyl compounds were not polymerized. Under polymerization of tri- and tetra-allyl monomers quaternary salts the cross-linked polymers are formed and under polymerization of diallyl monomer quaternary salts the water-soluble polymers of linear structure are formed. The absence of double bonds in macromolecules of such polymers was explained [5] by intra-molecular cyclization under chains growth during polymerization with formation of macromolecules consisted of regularly alternated cyclic units. At that the polymers on the base of N,N-diallyl-N,N-dialkylammoniumhalogenides (DADAAH) possessing the highest molecular mass in this group of polysalts were considered as the most perspective ones.

Cyclo-linear mechanism of chain propagation under DADAAH polymerization was suggested by Butler for the first time [6]. This mechanism [7] includes the series of intra- and intermolecular reactions:

1. addition of radical to double bond of monomer with formation of "open" radical;
2. intramolecular cyclization with formation of cyclic radical;
3. connection of cyclic radical with monomer molecule (Scheme 2.2).

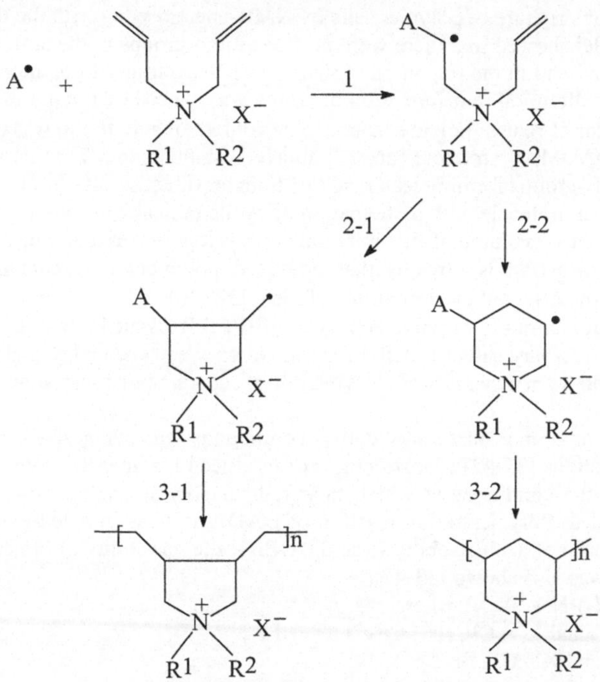

Scheme 2.2.

Such mechanism of cyclopolymerization of mentioned compounds was seemed to be very natural since it was based on the fact that activation energies of reactions of usual intermolecular chain propagation and intramolecular one differed slightly [8]. Tendency of such monomers to cyclopolymerization reactions was explained by Butler by lower energy expenses caused by electron interaction between reaction center (propagating radical) and the second double bond of monomer [9, 10].

Particularities of cyclopolymerization of various allyl monomers were studied in the number of works [11–18]. In these works they established the influence of some factors on proceeding of reaction of monomer's molecules connection to "open" radical, namely that reaction possibility was increased at high monomer concentrations [19, 20] and it was reduced in the presence of "bad" solvent [21] and also at process temperature decrease [22].

In the first works six-membered cyclo-linear structure of macromolecules was assigned to DADAAH polymerization products since such structure was considered as the most energetically efficient [5]. Later, as a result of improvement of methods of physical-chemical analysis, in particular NMR-spectroscopy more reliable data on the structure of poly-DADAAH were obtained and predominance of five-membered cycles in macromolecules structures was established [17, 23-25]. So, by comparison of ^{13}C NMR spectra of five-membered model poly-1,1,3,4-tetrapyrrolidine bromide and polydiallyldimethylammonium bromide the five-membered structure of polymer macromolecules structures was confirmed [24], and investigations of elementary stage of molecular chain propagation allowed establishing that the limitative stage of process was the interaction of "propagating" radical with the first double bond of molecule of diallyl quaternary salt, i.e. that stage of intermolecular growth. Then as a result of intramolecular cyclization under interaction of "internal" radical with the second double bond of monomer molecule the five-membered cyclic structure was formed.

The analysis of structure of polymer salts by NMR-spectroscopy with the use of resonance absorption of ^{14}N nuclei showed that there were no free amino-groups in the structure of PDADMAC [25]. The absorption band in the region of 1130cm^{-1} in IR-spectrum of polymer was explained [25] by the presence of sulfate-ion in it formation of which was caused by the use of ammonium persulfate (APS) as initiator of radical polymerization. The content of sulfate-ion was 0,4–0,8%.

Studies of PDADMAC structure [26, 27] allowed establishing [27] in its macromolecules the presence of end CH_3-groups forming as a result of transfer of active site from propagating polymer molecule to monomer molecule with participation of cyclic radical with five-membered structure.

At present the large volume of literature data on cyclopolymerization of diallyl compounds is accumulated from which it is obvious that produced polymers may contain as five-, so six-membered cyclic structures and even mixtures of them [28–36].

In the works of laboratory of polyelectrolytes of IPCS RAS the kinetics of radical polymerization of monomer quaternary salts of diallyl type in water, water-alcohol and alcohol mediums was studied in details and the mechanism of carrying out of separate polymerization studies was investigated.

A number of monomer quaternary diallyl compounds was synthesized in the laboratory for systematic investigations [37-42] (the structures of produced for investigations compounds are presented below) and the conditions at which their radical polymerization proceeded with essential rates were determined. Polymerization reaction of DADAAH was initiated by thermal destruction of APS or 2,2'-dinitrile of azobisisobutyric acid (AIBN); the possibility of effective photo-initiation of these monomer was also shown [43-47].

$(CH_2{=}CH\text{-}CH_2)_2N^+R^1R^2\,X^-$, where:

$R_1 = R_2 = CH_3$ and $X^- = Cl^-, Br^-, CF_3COO^-, BF_4^-$;

$R_1{=}R_2{=}C_2H_5$ and $X^-{=}\ Cl^-, Br^-$;

$R_1 = CH_3$; $R_2 = CH_2C({=}O)OCH(CH_3)_2$ and $X^- = Cl^-$

The kinetics of polymerization in water and water-methanol solutions of quaternary monomer salts ([**M**] = 0, –5,0 mole l^{-1}) was measured by dilatometer method at 50–75°C [40, 42] and also by densimetric method at 35°C [44, 45]. Studying of initial stage of polymerization at values of monomer conversion into polymer not higher than 3-5% showed that in all cases (even at ([**M**] = 5,0 mole l^{-1}) reaction was homogeneous and was completely inhibited by effective inhibitors – stable iminoxyle radicals. Spontaneous polymerization in the absence of radical initiator at experiment conditions was not observed.

We found that DADAAH polymerization rate in the wide range of monomers concentrations was characterized by half order of reaction by radical initiators (APS or AIBN) concentrations that testified to bimolecular mechanism of propagating chains termination and in given case it was definitely indicated to the absence of significant contribution of degradational chain transfer to monomer. The last fact distinguished this process from polymerization of monoallyl monomers for which polymerization reaction rate order by initiator was close to one due to the destruction of kinetic chains by DCT mechanism.

At the same time the investigations of dependence of kinetic orders of considered polymerization reactions by monomer at the rest equal conditions for a number of DADAAH gave surprising results. Dependences of initial rate of reaction of radical polymerization of monomers on their concentration deviated from linear dependence [44, 45]. The reaction is characterized by variable order by monomer exceeding the first one. Analogous data were obtained for DADMAC by Reinisch et. al. [46]. These results also testified to practically absolute absence of DCT to monomer (otherwise the rate order of reaction by monomer would be close to 0).

Authors of [47] investigated comparable contribution of elementary stages into total kinetics of DADAAH polymerization for clearing up of the reasons caused sharp and non-linear increase of initial rate of polymerization with the increase of monomer concentration. With the help of inhibitor method they showed that initiation rate was practically constant in the studied diapason of monomer concentration. In other words non-linear increase of polymerization rate was the result of corresponding increase of ratio of elementary constants of chain propagation and termination ($\mathbf{k_p} \, \mathbf{k_o}^{-1/2}$) [47, 48]. They also suggested that increase of monomer polymerization rate may be caused by anomalous increase of viscosity of its initial solution with the rise of monomer content [48] that led to the reduction of rate of elementary act of bimolecular chains termination. In accordance with this assumption taking into account North's ideas [49] and supposing that $\mathbf{k_o} \sim \boldsymbol{\eta}^{-1}$ the rate of DADAAH polymerization may be calculated by the equation:

$$\mathbf{V} = \mathbf{k_p} \times \mathbf{V_{in}}^{1/2} \times \boldsymbol{\eta}^{1/2} \times [\mathbf{M}] \qquad (1).$$

Experimental data confirmed the possibility of use of given equation for calculation of rates of polymerization of DADMAC and other monomer quaternary salts of diallyl type allowing revealing the influence of reaction solution viscosity on polymerization kinetics at small monomer conversion degree.

Authors of [47] concluded that the reason of found acceleration of DADAAH polymerization was the corresponding decrease of constant of rate of chains bimolecular termination $\mathbf{k_o}$ with the rise of monomer concentration due to diffusion control of this elementary stage beginning from the smallest conversions. They adduced principally important supposition about transformation of inactive allyl radical of molecular chain transfer into active radical of chain propagation at the expense of intramolecular reaction of cyclization. Cyclization of inactive allyl radicals occurs as a result of their connection by the second double bond of monomer.

The totality of these data indicates to the fact that radical formed as a result of act of chain transfer to monomer should be active enough. Appearance of "activity" of radical of chain transfer is connected with its transformation into another type of radical able to continue chain propagation in accordance with the mechanism suggested in the works [11, 18, 19] (see Scheme 2.3).

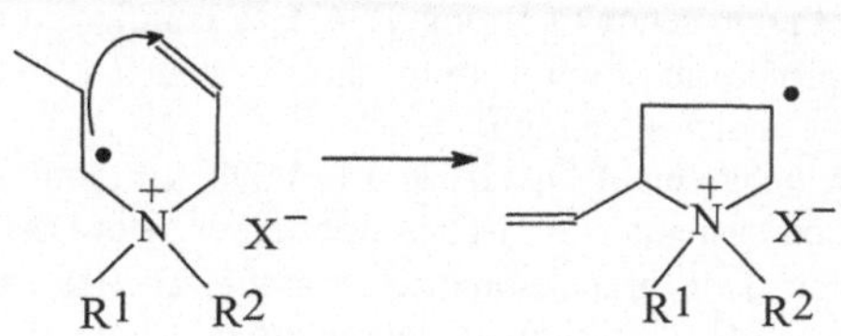

Scheme 2.3.

Found signals of end vinyl groups in the ^{1}H and ^{13}C NMR spectra of polymers are the evidences of transformation of degradational chain transfer into effective one [8, 11, 27]. Additional experimental data confirmed such mechanism of reaction [50–52].

2.1.2. Radical polymerization of monomers on the base of diallyl guanidine

In this Section we shall describe the main particularities and regularities of radical polymerization of monomer on the base of diallyl guanidine synthesized and investigated by the author and under his direction [53–58].

The radical polymerization of diallyl guanidine acetate was studied in water (redistillate), water-methanol and methanol solutions and radical initiators APS and AIBN were used as initiators ($[\mathbf{I}] = 10^{-2} - 10^{-3}$ mole l^{-1}) in the temperature interval 20–60°C. We established preliminarily that in the absence of initiator polymerization didn't occur.

Prepared reaction mixture was degassed in ampoule on the vacuum device (10^{-3} mercury), than ampoules were soldered up and placed in the thermostat. In the case when the initiator was decomposed at low temperatures (20°C, UV) reaction solution was replaced into quartz cuvettes (in vacuum).

On the base of experimental data we may conclude that homopolymerization of monomer salt DAGA at studied conditions is hindered. So, for example the conversion degree of monomer DAGA into polymer at conditions ([DAHA] = 4 mole l^{-1}; 60°C; [APS] = 4×10^{-3} mole l^{-1}; H_2O; polymerization time 72 hours) is about 5% ($[\eta]$ = 0,05 dl g^{-1}). All these facts testify to considerable contribution of degradational chain transfer to monomer under radical polymerization of DAGA monomer salt. At the same time, as it is known from Kabanov's et.al. [4] and Topchiev's et. al. [59-61] works introduction of inorganic complex formation agents or strong acids (as organic, so inorganic ones, including such acid as TFAA $\mathbf{pK_a}$ = 0,23) into diallyl monomers may considerably increase monomer's reactivity and inhibit DCT to monomer. They showed [61, 62] that as protonation, so quaternization of considered monomers led to the increase of activation energy of α-proton detachment reaction as a result of its bond strengthening due to which significant inhibition of degradational chain transfer to monomer was observed which was substituted by effective chain transfer [59, 61]. The last fact was confirmed by appearance in corresponding polymers' ^{1}H NMR spectra of protons' signals of end vinyl and methyl groups [27, 60, 61].

Taking into account these results the author with collaborators synthesized equimolar salt of diallyl guanidine trifluoroacetate (DAGTFA) for the first time (see Section 1.1) and its ability for radical polymerization was studied [54-58]. Polymerization of DAGTFA was carried out in water and methanol solutions at conditions of initiation by APS at 60°C and also photo-initiation at 20°C. We established preliminarily that polymerization reaction proceeded only in the presence of initiator. ^{1}H NMR and IR spectra of reaction solution were studied (60°C, $[\mathbf{M}]$ = 0,5 mole l^{-1}, polymerization time 20 hours). Carried out investigations showed that for given monomer as in the case of DAGA the rate of DAGTFA polymerization was not high and as a result of reaction polymer products with low molecular masses were formed that was confirmed by ^{1}H NMR method. So, in the considered case in spite of the presence of strong TFAA as counterion obviously enough strengthening of bond between α-proton and diallyl group didn't occur and probability of its detachment and consequently DCT to monomer were high.

We may explain such behavior of DAGA and DAGTFA monomer synthesized by us in the following way. In the studied systems cation-genic monomer salts DAGA and DAGTFA in water solutions may exist in the forms of three resonance structures (I–III), and the only one of them abbeys the condition necessary for inhibition of degradational chain transfer to monomer, i.e. the presence of charge on nitrogen connected with two allyl groups (structure I) (Scheme 2.4).

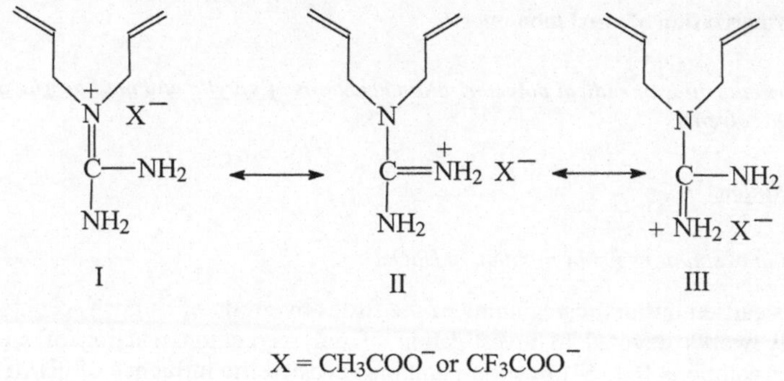

Scheme 2.4.

Whereas structures II and III most likely form stable (including spatial factors) delocalized systems with participation of two nitrogen atoms and acetate (or trifluoroacetate) counterion (IV) or at the expense of hydrogen connection of charged amino-group by delocalized carboxilate-ion (V) (Scheme 2.5).

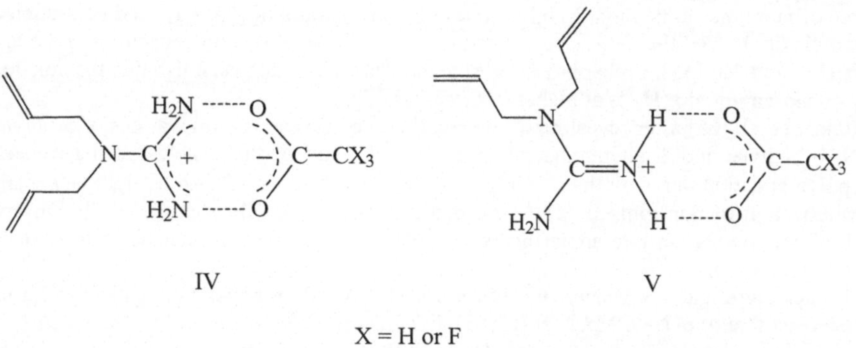

Scheme 2.5.

On the base of these suppositions we may conclude that under polymerization of monomer salts of DAG derivatives the degradational chain transfer to monomer will remain to a great extent. For the case when X = F (Scheme 2.5) we may expect reduction of the degree of DCT to monomer since high induction effect (-I) of trifluoromethyl group should lead to the reduction of stability of structures IV and V. However we didn't succeed in reaching of prospective effect and noticeable rates of DAGTFA polymerization possible due to insufficient solubility of monomer in studied systems.

Important data confirming our suggestions were obtained under analysis of IR-spectra of synthesized monomers and polymers. Analyzing the totality of results of NMR and IR-spectral investigations we established that for DAGA the most characteristic structure was structure IV, and for DAGTFA – the structure V (Sections 4.1 and 4.2.1).

Thus, under polymerization of DAGA and DAGTFA the main regulator of molecular mass value of synthesized polymer salts is degenerated chain transfer to monomer.

2.2. Radical polymerization of vinyl monomers

2.2.1. Kinetic particularities of radical polymerization reactions of acrylic and methacrylic acids in water and organic solvents

2.2.1.1. Water solutions

2.2.1.1.1. Variation of solutions pH by addition of NaOH

The works carried out in the beginning of the fifties by group of English scientists relate to the number of early works devoted to investigation of polymerization reaction of methacrylic acid (MAA) in water solutions [63, 64] in which authors revealed the influence of pH of reaction solution on general MAA polymerization rate where H_2O_2 was used as initiator. In mentioned works they established that in acid mediums at pH > 2 the rate of MAA polymerization was strongly reduced with the rise of pH (molecular masses of formed polymers were changed symbatic), and at pH > 5,5 MAA polymerization was not observed.

Since it is well known that in the studied diapason of pH the rise of monomer degree ionization occurs Kachal'sky and Blauer suggested [63, 64] that MAA monomer was able to participate in chains propagation reactions only in non-ionized form, i.e. "true" constant of the total polymerization rate $\mathbf{k_p}$ was connected with the observed propagation rate **k** by the ratio $\mathbf{k_p} = \mathbf{k}\ (1\text{-}\boldsymbol{\alpha})^{-1}$, where $\boldsymbol{\alpha}$ – degree of monomer ionization. In other investigations authors of [65, 66] used potassium persulfate as initiator showed that MAA polymerization in water solutions proceeded up to the high values of pH (about 13) [65]. Probably Kachal'sky and Blauer couldn't establish this fact due to inefficiency of chosen initiator H_2O_2 at higher (> 6) values of pH.

Pinner et. al. [66] after revealing of the fact that methacrylate-anion was easily copolymerized with N,N-diethylamino-2-ethylmethacrylate at pH = 7 came to the conclusion that reduction of MAA polymerization rate with the rise of pH in the interval of pH = 2,5–6,0 might be explained by copolymerization of non-ionized MAA and methacrylate-anion. They explained the observed reduction of polymerization rate under increase of pH by the increase of methacrylate-anion content in the initial mixture.

In later investigations Blauer [67, 68] measuring MAA polymerization kinetics initiated by AIBN showed that at pH > 6 MAA polymerization also proceeded and at that the reaction had the first order by monomer concentration (at pH = 6,0; 9,5; 11,0). On the base of this fact authors suggested that the first order by monomer was remained in the whole intervals of pH values. The order of reaction rate by initiator at pH = 10,6–11,6 was turned to be close to 0,5. In alkaline region of pH polymerization rate at first was slightly increased with the rise of pH and then it was reduced. Polymers molecular masses were changed in the same way. In these works they also established for the first time that under addition of low-molecular electrolytes (neutral salts) in the interval of pH = 6,8–9,6 polymerization rate was noticeable risen (Table 2.1).

For explanation of some increase of MAA polymerization rate in the alkaline region Blauer proposed that with the rise of pH the constant of rate of chains bimolecular termination was decreased due to electrostatic repulsion of charged macroradicals. If this supposition is right, then the rise of ion force of reaction solution under addition of low-molecular salts promoting collisions of likely charged particles should lead to the increase of bimolecular termination rate constant, i.e. to the decrease of polymerization total rate. However, on practice we observed the inverse phenomenon (Table 2.1).

Table 2.1. The dependence of total rate of polymerization of methacrylic acid on ion force of solution (61,4°C).

pH	Concentration, mole l^{-1}		$\mathbf{k} \times 10^5$, sec^{-1}
	monomer	salt (NaCl)	
6,8	0,2	-	1,19
7,0	0,2	0,4	1,78
9,5	0,1	-	1,51
9,6	0,1	0,4	2,00

Kinetics of polymerization of another weak unsaturated acid – acrylic (AA) – in water solutions was studied by Japanese researchers [69, 70]. And for this monomer the polymerization rate also depends on pH of reaction solution. Authors also showed that increase of ion force of reaction solution at pH = 7,2 as a result of addition of NaCl into it as well as in the case of MAA led to the rise of AA polymerization rate and increase of molecular masses of resulted polymers.

The most complete and systematic investigation of kinetic particularities and mechanism of radical polymerization reactions of AA and MAA in various mediums including water solutions was carried out by V.A. Kabanov, D.A. Topchiev et. al. [1, 3, 71-76]. As a result they established that in water solutions in the whole studied interval of pH (2,5–13,0) under the use of DAA as initiator as under AA, so under MAA polymerization (pH of solution was reached by addition of NaOH) the half order of the reaction rate by initiator concentration was remained that testified to bimolecular mechanism of chains termination. The first order by monomer concentration was observed under polymerization of MAA at pH > 2,5. And in the case of acrylic acid the order by monomer turned to be equal to 1,5.

The detailed investigation of kinetic particularities and regularities of radical polymerization of AA and MAA in water solutions showed that dependence of polymerization rates of mentioned monomers on pH of reaction solution had complex character. So, under polymerization of AA and MAA in water solutions in the interval of pH 2,5–7,0 considerable decrease of polymerization rate of corresponding monomers was observed. The values of resulted monomers molecular masses were changed cymbatic. In accordance with the values of ionization constants for AA $\mathbf{pk_a}$= 4,2, and for MAA $\mathbf{pk_a}$ = 4,32 (effective constants of PAA is $\mathbf{pk_a}$ = 6,4, for PMAA is $\mathbf{pk_a}$ = 7,0) we may conclude that in the region of pH < 6,0 with the rise of pH of reaction solution under polymerization of AA and MAA chain propagation occurs on non-charged macroradicals of polyacid. As it was shown in the region of pH > 6,0 the dependence of general polymerization rate of acrylate- and methacrylate-anions the rise of general rate of polymerization and polymers' molecular masses was observed. It is important that initiation rate in the interval of pH = 8,0–10,0 as in the presence, so in the absence of low-molecular non-polymerized salts is practically constant [73].

Thus, the observed increase of general rate of polymerization of acrylate- and methacrylate-anions with the rise of pH in alkaline region or under the increase of ion force at fixed pH was really caused by the increase of rates constants ratio of elementary reactions of chains propagation and termination, i.e. of the value $\mathbf{k_p}\ \mathbf{k_t}^{-1/2}$.

On the base of consideration of literature and experimental data V.A. Kabanov and D.A. Topchiev [1, 2, 72] set up a hypothesis about kinetic role of ion pairs under radical polymerization of ionized monomers. For the cases of acrylate- and methacylate-anions polymerization in water solutions this hypothesis was formulated and grounded in the following way. In water solutions at pH > AA and MAA were completely ionized, i.e. in the reaction solution there were only acrylate- and methacrylate-anions. PAA and PMAA are weaker than corresponding monomer acids and their ionization with the rise of pH occurs in alkaline region, i.e. the observed in this region of pH kinetic effects are accompanied by the change of chemical nature of propagating chains. The possibility of change of effective reactivity of macroradicals was considered by given hypothesis. They suggested that if ionized, i.e. negative charged macroradicals of propagation were able to form ion pairs with low-molecular cations (Cat^+, Scheme 2.6) present in solution in particular at the ends of propagating chains, then it should lead to the rise of chain propagation reaction rate due to relief of electrostatic repulsion between propagating radical and likely charged monomer. Then the electrostatic repulsion under connecting of monomer anions with likely charged radicals in chain propagation acts should be leveled.

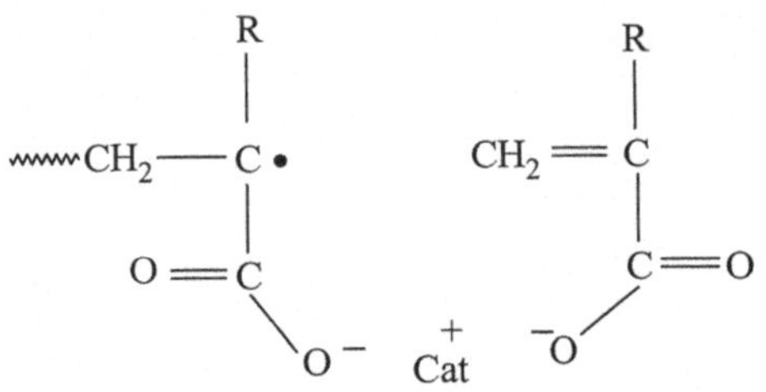

Scheme 2.6.

The probability of formation of ion pairs at the ends of propagating chains should be increased with the rise of pH that in the framework of hypothesis should lead to the increase of effective constant of propagation rate $\mathbf{k_p}$. Actually [73], increase of Na^+ ions concentration at given monomer concentration and fixed value of solution pH (pH > 7,0) is accompanied by the rise of polymerization general rate. (The notion "ion pair" is used as Fuoss's term, i.e. we suggested that there is direct contact of solvate capsule of oppositely charged ions [1]).

It is well known that polyanions of polyacids are able to connect low-molecular counter-ions in water solutions [77]. Usually the affinity between oppositely charged low-molecular ions, for example between Na^+ and CH_3COO^- in low-concentrated water solutions is insufficient for stabilization of ion pairs. Formation of stable ion pairs is observed only in less polar organic solvents. However in the case of polyelectrolytes the electrostatic field of the whole poly-ion promotes connecting. That is why at high ionization degrees significant part of aonter0ions turned to be immobilized in polymer balls.

For reception of direct information on the influence of pH and ion force of solution on constants of rates of elementary stages of process in work [73] they carried out the detailed kinetic study of methacrylate- and acrylate-anions photo-polymerization in water solutions which pH was fixed by NaOH. The results of carried out investigation and measurements are presented in Tables 2.2 and 2.3.

Table 2.2. The constants of rates of elementary reactions under polymerization of acrylate-anion ([**M**] = 1,2 mole/l; [AIBN] = $8{,}3\times10^{-4}$ mole l^{-1}; 23°C)

pH	$\mathbf{V}\times10^6$, mole l^{-1} sec^{-1}	$\mathbf{V_{in}}\times10^9$, mole l^{-1} sec^{-1}	$\mathbf{k_p\, k_t^{-1/2}}$, $mole^{1/2}$ $l^{-1/2}$ $sec^{-1/2}$	$\mathbf{k_t}\times10^8$, mole l^{-1} sec^{-1}	$\mathbf{k_p}$, mole l^{-1} sec^{-1}
7,9	5,1	9,1	0,04	2,6	650
7,9*	25,0	9,1	0,20	2,6	3100
11,0	51,0	9,1	0,40	2,7	6600
13,6	21,0	9,1	0,17	2,8	2500

Notes: * – 1,5n solution of NaCl in water.

Table 2.3. The constants of rates of elementary reactions under polymerization of methacrylate-anion ([**M**] = 0,92 mole/l; [AIBN] = $2{,}5\times10^{-4}$ mole l^{-1}; 23°C)

pH	$\mathbf{V}\times10^6$, mole l^{-1} sec^{-1}	$\mathbf{V_{in}}\times10^9$, mole l^{-1} sec^{-1}	$\mathbf{k_p\, k_t^{-1/2}}$, $mole^{1/2}$ $l^{-1/2}$ $sec^{-1/2}$	$\mathbf{k_t}\times10^8$, mole l^{-1} sec^{-1}	$\mathbf{k_p}$, mole l^{-1} sec^{-1}
8,0	4,2	8,9	0,046	2,1	670
13,6	12,0	8,9	0,130	2,3	1900

As it is obvious from presented data with the rise of pH from 8 to 13,6 propagation rate constant $\mathbf{k_p}$ is increased as much as stationary rate of process. Termination rate constant at that is practically not changed, i.e. according to the hypothesis the rise of polymerization rate of acrylate- and methacrylate-anions polymerization in alkaline region of pH is actually caused by the increase of propagation rate constant $\mathbf{k_p}$.

Practical constancy of bimolecular termination rates constants with the rise of pH and extension of solution ion force in alkaline region establishing under polymerization of acrylate- and methacrylate-anions in accordance with authors opinion may be explained only if they suppose that termination is limited by diffusion processes. Since the hydrodynamic sizes and consequently the rate of forward diffusion of balls themselves are significantly changed in the studied intervals of reaction medium properties change we may propose that the limiting stage is the diffusion to each other of two end units of macromolecular balls which according to North's theory [49, 78] preliminary coming together to the distance less than critical.

In the acid region of pH (pH < 6) under AA and MAA polymerization non-charged macroradicals of polyacids carry on chains propagation and with the rise of pH the content of ionized monomer – acrylate- and methacrylate-anion correspondingly is increased. For revealing of reasons of observed sharp reduction of general rate of MAA polymerization with the rise of pH they estimated the relative change of chain propagation rate constant (they used the method of chain transfer in the presence of mercaptoethanol) and showed that established reduction of general rate of polymerization was caused by the decrease of rate constant of chain propagation elementary reaction $\mathbf{k_p}$ [76].

As a result of measurements of kinematic viscosity of model solutions they found [76] that under the change of pH in the interval of values 2,5–5,5 hydrodynamic sizes of polymer chains and consequently of macroradicals were not significantly changed, so the conformational state of propagating chains in considered diapason of pH wasn't changed. At the same time, with the rise of pH at studied conditions the ionization of MAA was observed and in solution the ratio of two in essence different monomers methacrylic acid and methacrylate-anion able to connect to non-ionized radicals of PMAA propagation was changed. It is known that MAA and methacrylate-anion are characterized by various values of parameters of reactivity **Q** and **e**. Since methacrylate-anion concentration in the studied diapason of pH is increased with the rise of pH and considering (in accordance with the value of constant of MAA ionization) that at pH = 2,5 MAA is completely non-ionized and at pH = 5,5 in reaction solution there is only methacrylate-anion we may calculate the ratio of constants of rates of MAA and methacrylate-anion connection to non-ionized radical of propagation in the framework of Q-e scheme:

$$\mathbf{k'_p / k_p'' = Q_1 \cdot exp(-e_1^2) / Q_2(-e_1 \cdot e_2)} \approx 0,5,$$

where $\mathbf{Q_1} = 2,34$; $\mathbf{e_1} = 0,65$ – the constants for MAA; $\mathbf{Q_2} = 1,36$; $\mathbf{e_2} = -1,18$ – the constants for methacrylate-anion.

Actually, according to [76] the ratio of constants is equal to 0,5 and MAA polymerization rate in the interval of pH = 2,5–5,5 is decreased in 5 times, i.e. experimental result couldn't be explained only by differences in classical parameters of "comonomers" reaction ability.

Obviously, the most probable reason of observed sharp reduction of chains propagation rate constants with the rise of pH from 2 to 6 as authors of [1, 3, 76] think is caused by the fact that with MAA ionization the concentration of strongly hydrated methacrylate-anions in relatively hydrophobic non-charged balls of PMAA macroradicals turn to be lower than their medium concentration in solution. At the same time in acid mediums we may expect predominant sorption of MAA in PMAA balls, i.e. the rise of monomer concentration in micro-regions surrounding active sites.

2.2.1.1.2. Regulation of solutions pH by addition of amines

The data on studying of influence of various neutralization agents including organic amines which basicity and structure may be widely varied on polymerization process of considered monomers are of great interest.

The main results of investigation of such systems were obtained in Kabanov's and Topchiev's et. al. works [1, 3, 71-75]. Isobutylamine (IBA), ethylene diamine (EDA), triethylamine (TEA) and ammonium hydroxide were chosen as neutralizing agents under carrying out of kinetic investigations of MAA polymerization; under AA polymerization TEA and NH_4OH were used as neutralizing agents. They established that in all mentioned systems the half order by initiator (AIBN) was maintained. Under MAA polymerization the first order by monomer was determined, and under AA polymerization there was the sesquialteral order. They also showed that the rate of AIBN initiator destruction under MAA polymerization didn't depend on nature of neutralizing agent and was practically constant in the investigated interval of pH values of reaction solutions [73].

Considering the found earlier particularity of behavior of acrylic acid which in contrast to MAA is characterized by relatively high ability to connect primary and secondary organic amines by Michael'e reaction under the studying of AA polymerization in water in dependence on pH the TEA was chosen which effect on kinetics of polymerization was compared with ammonia effect [1, 3, 73].

The course of AA and MAA polymerization dependence on reaction solution pH has complex character. In the region of pH = 2,0-4,5 the nature of neutralizing agents practically don't influence on polymerization rate and their role according to [1] is come to monomer ionization. With the rise of pH the methacrylate(acrylate)anion content is increased (the reasons of the decrease of AA and MAA polymerization rate with the rise of pH are discussed above), and propagating polymer chains in the considered interval of pH are not ionized.

At high pH (5–10) the dependence of MAA polymerization rate in the case of addition of EDA or NH_4OH into reaction system passes through the extremum. For the system MAA–EDA the maximum rate is observed at pH = 6,6 when macroradicals of PMAA are preliminary not ionized (EDA dissociation constants are $\mathbf{pK_{a1}} = 7{,}0$; $\mathbf{pK_{a2}} = 10{,}0$). Authors proposed that non-ionized amino-groups of EDA presenting in reaction solutions at given conditions were able probably to form associates with non-ionized carboxyl groups at the ends of propagating chains. In such case transition state in the act of chain propagation for given system may be presented as follows (Scheme 2.7).

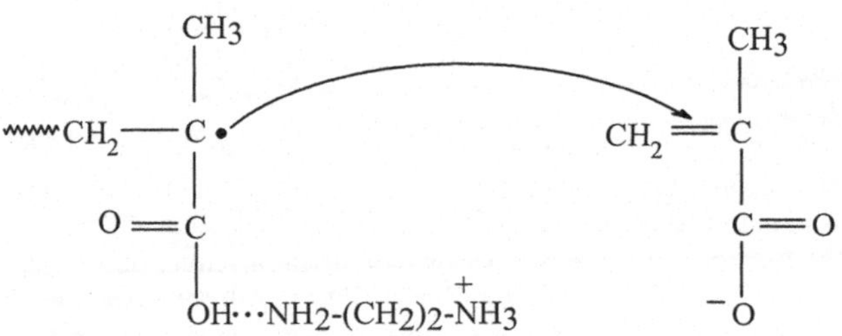

Scheme 2.7.

That is why we may expect that electrostatic attraction of protonated (second) amino-group of EDA and methacrylate-anion should additionally promote connection of methacrylate to active site connected by hydrogen bond with non-ionized aminogroup of EDA. Hydrophobic interactions make significant contribution into stabilization of such associative structures. However, when passing to neutral pH (6,8–7,5) concentration of such associates is obviously reduced (due to ionization of carboxyl groups in the units of propagating chains) that leads to the decrease of polymerization rate.

In alkaline region of pH the presence of EDA and IBA (pK = 10,4) leads to sharper increase of methacrylate-anion polymerization rate with the rise of pH than under addition of NaOH. In the framework of hypothesis about kinetic role of ion pairs one may explain this fact in the following way: at pH > 7,5 when macroradicals are preliminary ionized at the ends of propagating chains the ion pairs are formed with participation of carboxylate-anions of macroradicals and organic cations and this promotes chain propagation acceleration. Obviously in such cases the hydrophobic interactions additionally stabilizing ion pairs promote also formation of ion pairs with participation of ethylendiammonium and isobutylammonium.

Probably, hydrophobic effects are responsible also for observing essential differences in kinetics of polymerization of acrylate- and methacrylate-anions in water solutions which pH was set by addition of triethylamine. Significantly higher kinetic effect with the rise of pH under methacrylate-anion polymerization was explained by the authors by earlier established fact that for PMAA chains hydrophobic interactions were significantly higher than for PAA [79]. They also suggested that namely such interactions additionally stabilize ion pairs at the ends of ionized radicals of PMAA propagation.

Interesting results were obtained in the course of kinetic investigations of polymerization of considering acids under the use of weak inorganic base NH_4OH as neutralizing agent. Authors found that as under polymerization of AA, so under polymerization of MAA sharper (in comparison with addition of NaOH) rise of polymerization rate was observed. This increase was especially noticeable under polymerization of methacrylate-anion.

The established differences in MAA and AA polymerization rates in the region of pH = 6-9 in the presence of NaOH and NH_4OH authors of [75] explained by specificity of connection of Na^+ and NH_4^+ ions with PMAA that was obvious form consideration of titration curves of corresponding polyacids [3]. Significantly higher kinetic effect (with the rise of pH) under polymerization of methacrylate-anion was also explained by pronounced hydrophobic properties of PMAA (in comparison with PAA). They assumed that in the case of NH_4OH addition into reaction solutions as well as in the cases of EDA and IBA the hydrophobic interactions should additionally stabilize ion pairs at the ends of ionized radicals of PMAA.

Authors of [75] also found qualitatively different type of neutralizing agent influence under MAA polymerization in water mediums when pH of reaction solution was set by addition of weak organic base pyridine (Py) (pK_a – 5,19). Methacrylic acid polymerization dependence on pH in the presence of pyridine-ions is significantly differed from analogous dependence found in NaOH solution and other studied neutralizing agents – EDA, IBA, NH_4OH. In all mentioned cases they observed the decrease of polymerization rate with the rise of pH up to 4,5.

The characteristic feature of established dependence is the presence of maximum in the interval pH = 3,5–4,0, and reaction rate turns to be by order higher than methacrylate polymerization rate in NaOH solution at the same values of pH. The rate is increased even in the cases when there is relatively small amount of pyridine. And initial molar ratio [Py] : [MAA] = 0,23 corresponds to maximum polymerization rate.

Authors considered probable influence of presenting in water solutions linear dimmers of MAA which were reported in some works [80, 81] as possible reasons explaining anomalous course of MAA polymerization rate curve. They also established that in water solutions of carbonic acids there were linear dimmers (Scheme 2.8) whereas in non-polar solvents there were cyclic dimmers [82].

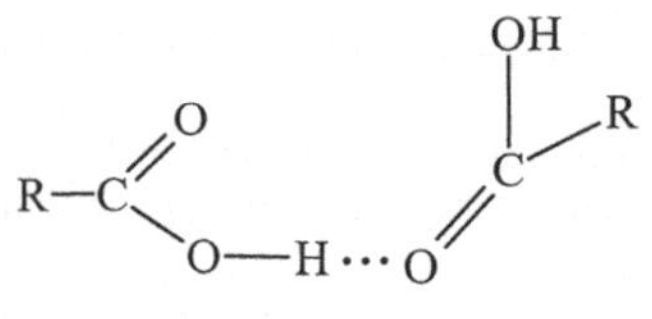

Scheme 2.8.

The rise of MAA polymerization rate at pH = 3–4 accompanying by blooming of the system and polymer precipitation is caused by formation of donor-acceptor complex between ionized dimmer of MAA (in which the system of conjugated bonds plays role of electrons donor) and pyridine ion (electrons acceptor) which is stabilized by hydrophobic interactions (Scheme 2.9).

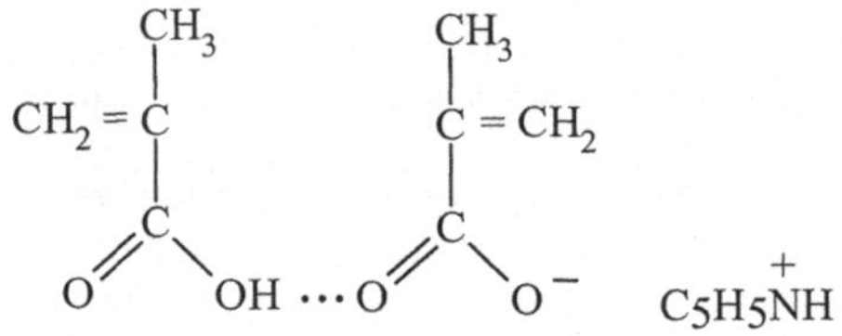

Scheme 2.8.

In the formed complex redistribution of electron density on double bonds occurs that may lead to chemical activation of dimmer. Increase of solution pH up to pH = 5–6 leads to ionization of the second molecule of monomer entering into dimmer and consequently leads to complex destruction and sharp polymerization rate decrease. Ionization constants of such dimmers are higher than ionization constants of corresponding monomer acids. Hydrophobic interactions of alkyl groups play the main role in stabilization of linear dimmers [81].

The role of hydrophobic interactions in considering process was shown also while investigating polymerization kinetics of another unsaturated acid AA in water solutions which pH were set by addition of NaOH and pyridine. It is known, that hydrophobicity of AA is feebly marked than the hydrophobicity of MAA. It turned to be that change of AA polymerization rate in the interval pH = 2–6 weakly depended on the nature of neutralizing agents, i.e. in the presence of pyridine-ions acceleration of AA polymerization wasn't observed.

As a result of carried out investigations of given phenomenon authors concluded that in studied system the case of interaction of self-regulating activator pyridine-ion with dimmer of MAA was realized as a result of which the complex with higher reactivity than for MAA was formed.

Thus, considered in given section results of investigation of methacrylate- and acrylate-anions polymerization kinetics in water solutions of bases and organic amines testify to significant influence of neutralizing agent nature on dependence of initial rate of non-saturated acids polymerization on pH of reaction medium. In accordance with proposed hypothesis authors of [1] considered this phenomenon as a result of specific influence of ion pairs nature on chain propagation rate constant under polymerization and on propagating radical of ionized polyacid-polycation, or they also explained it by formation of complex of ionized monomers in water mediums (for example, the system MAA–pyridine).

2.2.1.2. Polymerization of unsaturated acids in water and organic mediums

The basis of kinetic regularities and particularities of radical polymerization of monomers of acryl type in water solutions in dependence on the change of pH in the presence of various neutralizing agents was considered in the section 2.2.1.1. It seems to be very important to estimate also the influence of nature of reaction medium on polymerization process of considering monomers, in particular to make comparable analysis of kinetic data under polymerization of acrylic acids in organic solvents and water solutions. a number of publications is devoted to this problems [83–89].

It is known that under radical polymerization of unsaturated acids in organic solvents the electron-donor ability of solvents and consequently the degree of monomer solvation in reaction system plays an important role. For dimmer form of unsaturated acids the high value of $\mathbf{Q_1}$ is characteristic, i.e. higher reaction ability in the framework of scheme **Q–e** since dimmer form is characterized by higher energy of double bond stabilization in comparison with monomer one.

The most detailed quantitative investigations of activating role of water and H-ions in process of unsaturated acids and amides polymerization were made by Abkin et. al. [83–89]. In presented works the authors studied AA, MAA and fluoroacrylic acids (FAA) polymerization initiated by photochemical destruction of AIBN in the following solvents: water, dimethyl formamide (DMFA), dimethyl sulfoxide (DMSO).

It follows from Table 2.4 that under transition from water solutions to DMFA ones and especially DMSO the chain propagation rate constant $\mathbf{k_p}$ and chain termination constant $\mathbf{k_t}$ are noticeably reduced (rate of polymerization is also reduced).

Table 2.4. Parameters characterizing reactions of propagation and termination of chains under polymerization of MAA and AA in various solvents (UV λ = 365 nm, 20°C)

Solvent	Monomer	$\mathbf{k_p}\times10^3$, l mole^{-1} sec^{-1}	$\mathbf{E_p}$, kkal· mole^{-1}	$\mathbf{A_p}\times10^7$, l mole^{-1} sec^{-1}	$\mathbf{k_o}\times10^7$, l mole^{-1} sec^{-1}	$\mathbf{E_o}$, kkal mole^{-1}	$\mathbf{A_o}\times10^7$, l mole^{-1} sec^{-1}
H_2O	AA	27,2	3,1	0,6	18	0	18
	FAA	26,0	4,5	0,6	8,7	0,6	20
	MAA	4,1	4,3	0,67	1,1	0,2	1,6
DMFA	AA	4,2	5,6	6,2	10	0	10
	FAA	2,2	4,9	0,75	7,4	0,4	15
DMSO	AA	0,5	8,0	47	2,0	0	2,0
	FAA	0,85	5,2	0,66	4,9	0,4	10,0
	MAA	0,15	7,6	4,1	0,9	1,0	4,8

They also established that presented phenomenon is not connected with the rise of dielectric permeability of reaction system [90] since at equal water concentration the polymerization rates in various mediums are close however dielectric permeability is varied in wide limits.

Let us note that the most significant change of propagation constant $\mathbf{k_p}$ is observed under transition from water to DAMSO for AA (in 50 times) and less significant for FAA (in 30 times). As the authors of these works think the found influence of solvents nature on radical polymerization kinetics of considering unsaturated acids is connected with ability of concrete solvent to destruct dimmers of monomer acids and in that way to change their reaction ability. However this supposition seemed to be insufficiently grounded since as it is known content of dimmerized AA and DMSO molecules doesn't exceed 10% (according to cryoscopic measurements). In the water AA exists in the state of separate molecules (concentration of dimmer form is insignificant), as well as FAA which doesn't reveal their inclination to associative interactions.

That is why probably the observed kinetic effects under polymerization of acrylic acids in various solvents couldn't be explained only by formation of dimmers between molecules of acrylic acids. The influence of solvent nature may be connected with the number of reasons among which the following ones should be noted:

1. formation of donor-acceptor complexes between polymer radicals and solvents molecules leading to the reduction of radicals reaction ability;
2. solvation of monomer molecules and polymer radicals or specific interaction of these particles with solvent molecules in particular at the expense of formation of hydrogen bonds leading to

the change of electrons density on C=C-bonds of monomer or non-coupled electron in radical and consequently to the change of reactivity of reacted particles;

3. change of conformational characteristics of polymer molecules in solvents possessing various solubilizing ability in relation to polymer.

Besides distinctive particularity of water-soluble monomers polymerization is hydrophobic interaction appearing in water solutions of polymers. It leads to significant difference in kinetic parameters of polymerization of this class of monomers in water and organic solvents solutions. Hydrophobic interaction also causes significant reduction of polymerization rate and value of $\mathbf{k_p}$ under addition to water of small amounts of organic solvent due to preferred sorption of organic solvent by polymer ball. Naturally formation of inter-chain associates as a result of hydrophobic interactions and consequently their "catalytic" influence should depend on hydrophobicity of used monomers.

Authors of considered works [83–89] also think that definite contribution into observing kinetic effects are made by factors connected with conformational state of propagation chains in mentioned solvents (i.e. influence of change of conformational state of macroradical under the change of solvent nature on kinetic of process due to the change of accessibility of polymerizing particles for active sites). Assuming in principle the possibility of contribution of this factor into polymerization kinetics in considered systems we should draw your attention to the following. As it was mentioned above for water solutions (Table 2.4) higher constants of chains termination and propagation are characteristic (and higher rates of polymerization also). The same effects were observed for some other water-soluble monomers. That is why we may suppose that the reason of this may be greater solvation (in comparison with organic solvents), i.e. hydration of polymerized particles in water mediums. However, of course various monomers may be characterized by various degree of solvation. We also may assume that in considered number of monomers especially for AA the inclination of polymerized particles – monomer molecules and macroradicals to hydration in water solutions should be observed. That is why it is not surprising that under transition from DMSO to water the highest kinetic effects are found under AA polymerization. Let us also note that while discussing the question about comparable reaction ability of considered monomers in various solvents it is useful to have corresponding data on copolymerization ability of these monomers.

Interesting results were obtained by Shapiro et. al. under investigation of hydrogen bond at conditions of hetero-phase polymerization of AA and MAA when polymerization was initiated radiationally chemically [91–93]. It turned out that solvents able to form hydrogen bond – dioxane, acetic acid, methanol, water – slightly influence on AA polymerization rate in hetero-phase region. Whereas hydrocarbon solvents –toluene, hexane – being precipators in relation to PAA lead to sharp reduction of AA polymerization rate and decrease of length of formed chains. Under MAA polymerization the effects mentioned above are revealed to a significantly less degree.

Authors of [91–93] proposed that solvents of the first group increase life-time of AA linear associates cross-linking them, and solvents of the second group (hydrocarbon ones) shift equilibrium toward dimmer form. However, in contrast to authors' opinion it is hard to suppose that solvents able to form H-bond and participate in mixed associates don't destruct but strengthen liner associates of AA. It is also hard to expect that decrease of relative amount of AA linear associates is enough for reduction of AA polymerization rate almost by order. That is why we think that it is most likely to explain found kinetic effects in terms of the change of constants of bimolecular and monomolecular termination of propagating chains.

Thus, the results presented above bring out clearly that the nature of solvent is the very important factor determining the course of radical polymerization of water-soluble polymers. The influence of the solvent in the case of water-soluble monomers is significantly higher than in the case of radical polymerization of such non-polar monomers as styrene for example. Especially high change of both total rate of reaction and value of $\mathbf{k_p}$ is observed under transition from polymerization in organic solvents to polymerization in water solution.

2.2.2. Radical polymerization of acrylate- and methacrylate guanidines in water mediums

Salts of the acids of acrylic type containing chemically active functional groups undoubtedly represent the perspective type of monomers since polymers and copolymers produced on their base may safe potential of activity being convenient carriers of biologically active substances also. Requirement of such products in various branches of industry beginning from technique and ending with medicine naturally stimulates and undoubtedly justifies the necessity of study of particularities of synthesis processes and formation mechanism of given class of polymer compounds.

In given section of the book the results are summarized obtained by authors and under his direction while studying reactions of radical polymerization of firstly synthesized monomer salts on the base of acrylic, methacrylic acids and guanidine (GA, MAG) and the characteristic particularities and basic kinetic regularities of given process are revealed. These results are partially published in [94–98].

Since all kinetic particularities were carried out with the use of dilatometry method, so we preliminarily determined the specific concentrations of polymers PAG and PMAG produced under radical polymerization of corresponding monomers in water mediums. Their values were equal to 10,8% for PAG and 7,0% for PMAG.

Carried out systematic investigations showed that GA and MAG polymerization reaction at conditions of kinetic measurements proceeded only in the presence of radical initiators (APS and AIBN), "spontaneous" polymerization (in the absence of initiator) was not observed and polymerization reaction was completely hindered under addition into initial reaction medium of effective radical inhibitor – stable iminoxyl radical 2,2,6,6-tetramethyl-4-hydroxy-1-oxyl. The values of polymerization rates were determined by initial linear section of kinetic curve at the conversion not higher than ~ 5%.

2.2.2.1. Particularities of radical polymerization reactions of acrylate- and methacrylate guanidines

Carried out kinetic investigations of radical polymerization of GA and MAG in water and organic mediums showed that polymerization processes of mentioned monomers were characterized by the number of specific particularities. In all organic solvents (methanol, ethanol, dioxane, initiator AIBN) GA and MAG polymerization is heterogeneous. The appearance of white flaky sediment in reaction volume of dilatometer beginning from the initial (practically from zero) conversion testifies to the last fact.

Under GA and MAG polymerization in water solutions the homogeneity (complete transparence of the solution) up to high conversions **q** ~ 50–70% remains only in definite concentration interval of monomers which is equal for GA $0{,}25 \le [\mathbf{M}] \le 1{,}30$ mole l^{-1} and for MAG $0{,}20 \le [\mathbf{M}] \le 0{,}40$ mole l^{-1}. In more concentrated water solutions of considered monomers even at low conversions **q** ~ 3 – 5% the turbidity of reaction medium is observed (without phase separation). We proposed that the probable reason of found effect was formation in water solution in the course of GA and MAG polymerization of associative structures stabilized by hydrogen bonds and (or) hydrophobic interactions. Since as it is known [1] during the polymerization process of ionogenic monomers the influence of a number of factors (dissociation, specific and non-specific bonding of ions, electrostatic and hydrophobic interactions, etc.) on the state of polyelectrolytes chains in solution was possible it was naturally to expect the revealing of analogous effects also during the propagation of PAG and PMAG chains. Possible reasons of observing heterogeneity will be considered in details below (Section 2.2.2.2).

It is important to note that synthesized polymers PAG and PMAG are insoluble and even don't swell in any f investigated by us organic solvents – methanol, ethanol, dioxane, DMSO, DMFA. At the same time we established that when the values of synthesized polymers intrinsic viscosity were high enough – $[\eta] \approx 0{,}1–2\ 5$ dl g^{-1} for PMAG and $[\eta] \approx 0{,}1–0{,}94$ dl g^{-1} for PAG they were easily solved in water that indicated on their linearity. The data of element analysis of polymers confirmed that their structure corresponds to theoretical.

Actually, prepared separately solutions of GA and MAG monomers and PAG and PMAG polymers are absolutely transparent, but under their mixing (as under heating, so at room temperature) turbidity of mixture is observed that is caused as we propose by specific interactions between monomer and polymer chain analogous to polymer "salting-out".

As kinetic investigations showed micro-heterogeneity of process (i.e. appearance of opalescence) was observed by sight at the following monomers concentrations: [GA] > 1,3 mole l^{-1} and [MAG] > 0,4 mole l^{-1}, at **q** ≤ 5%. At that nevertheless as at conditions of complete homogeneity of system under GA polymerization, so at micro-heterogeneity under AG and MAG polymerization the half order of reaction by initiator (APS) concentration is maintained, as it is obvious from (Figure 2.1).

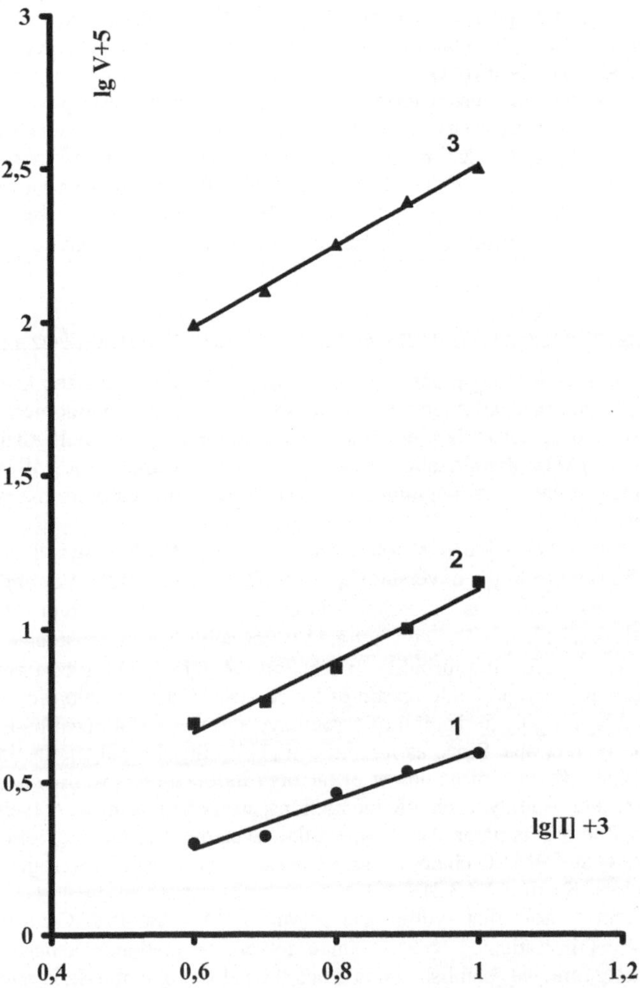

Figure 2.1. The dependence of GA and MAG polymerization initial rate on concentration of initiator APS; **1** – [MAG] = 1,0 mole·l^{-1}; **2** – [GA] = 1,0 mole·l^{-1}; **3** – [GA] = 1,5 mole·l^{-1}; H_2O;' 60°C.

We also should note that preliminary before the beginning of systematic kinetic investigations we studied stability of GA and MAG monomers to hydrolysis with the help of NMR spectroscopy method at polymerization reaction conditions. We showed that under heating of initial reaction solution for 20 hours (60°C, as with, so without hydroquinone) the monomers were stable (NMR spectra of initial reaction solutions of both monomers before and after heating were absolutely identical). The main kinetic measurements of GA and MAG polymerization in water solutions were carried out under the use of APS initiator due to its good solubility and efficiency in studied reaction solutions.

Investigation of kinetic orders of considered polymerization reactions by monomer showed that with the increase of initial GA and MAG concentrations the initial polymerization rate was increased sharply and non-linearly and reaction was characterized by exceeding the first variable order by monomer (Figures 2.2 and 2.3).

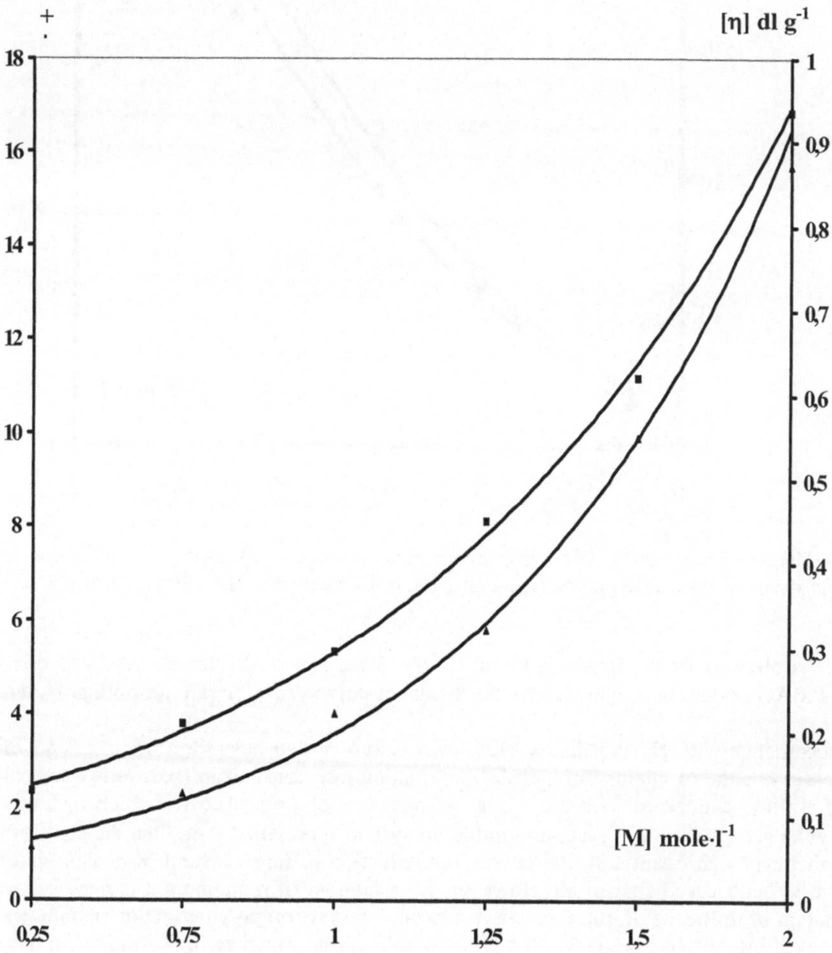

Figure 2.2. The dependence of GA polymerization initial rate (curve **1**) ([APS] = 5×10^{-3} mole l^{-1}; 60°C) and intrinsic viscosity of resulted polymers (curve **2**) (1n NaCl solution in H_2O; 30°C) on monomer concentration.

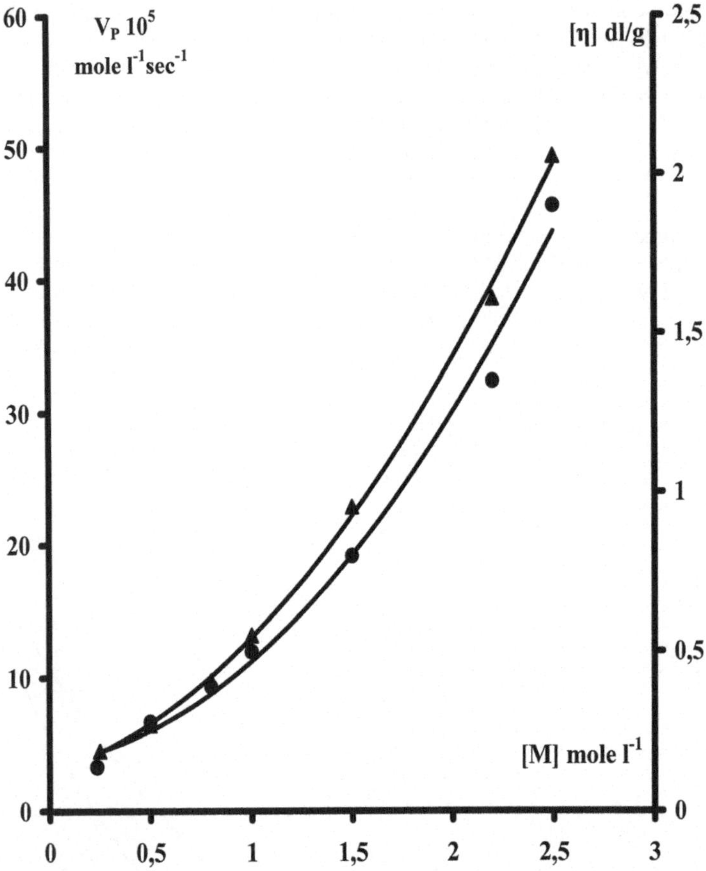

Figure 2.3. The dependence of MAG polymerization initial rate (curve **1**) ([APS] = 5×10^{-3} mole l^{-1}; 60°C) and intrinsic viscosity of resulted polymers (curve **2**) (1n NaCl solution in H_2O; 30°C) on monomer concentration.

As it is obvious from Figures 2.2 and 2.3 the dependence of intrinsic viscosity of resulted PAG and PMAG polymers is cymbatic to the change of polymerization rate for both polymerization systems.

In accordance with North's theory [49, 78] it is known that observing in radical polymerization processes change of chains bimolecular termination rate constant $\mathbf{k_t}$ (reaction is controlled by diffusion) is often connected with the change of reaction solution viscosity which is naturally increased by the accumulation of reaction product in system – polymer. And then the contribution of viscosity factor is significant and that is why the reduction of rate constant of chains bimolecular termination $\mathbf{k_t}$ is observed first of all. However, for a number of monomers it is necessary to consider the factor of influence of initial reaction solution viscosity on polymerization parameters.

For revealing of reasons causing the found non-linear dependence of $\mathbf{V_p}$ on **[M]** in our case it was interesting to clear up the character of dependence of macroscopic viscosity of initial solutions of investigated monomers on their concentration (Figures 2.4, 2.5).

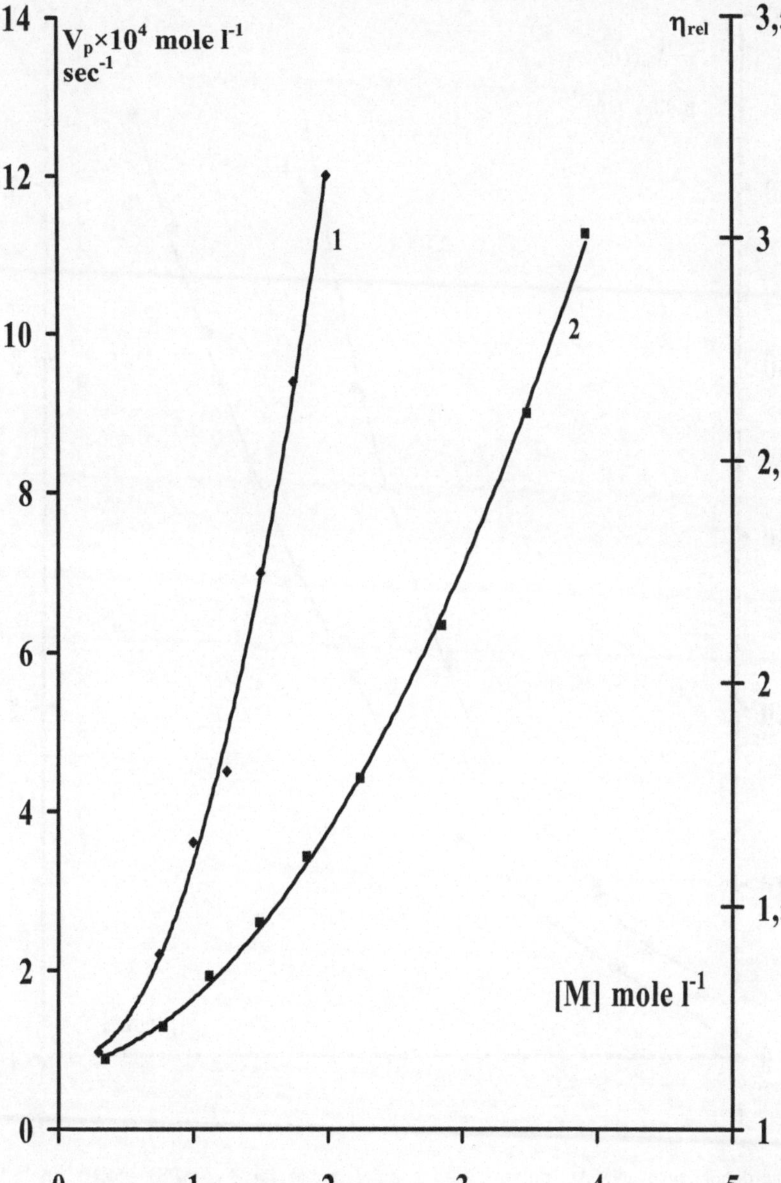

Figure 2.4. The dependence of GA polymerization initial rate (curve **1**) ([APS] = 5×10^{-3} mole l^{-1}; 60°C) and relative viscosity of initial GA solutions (curve **2**) (1% hydroquinone; 30°C) on monomer concentration in water solutions.

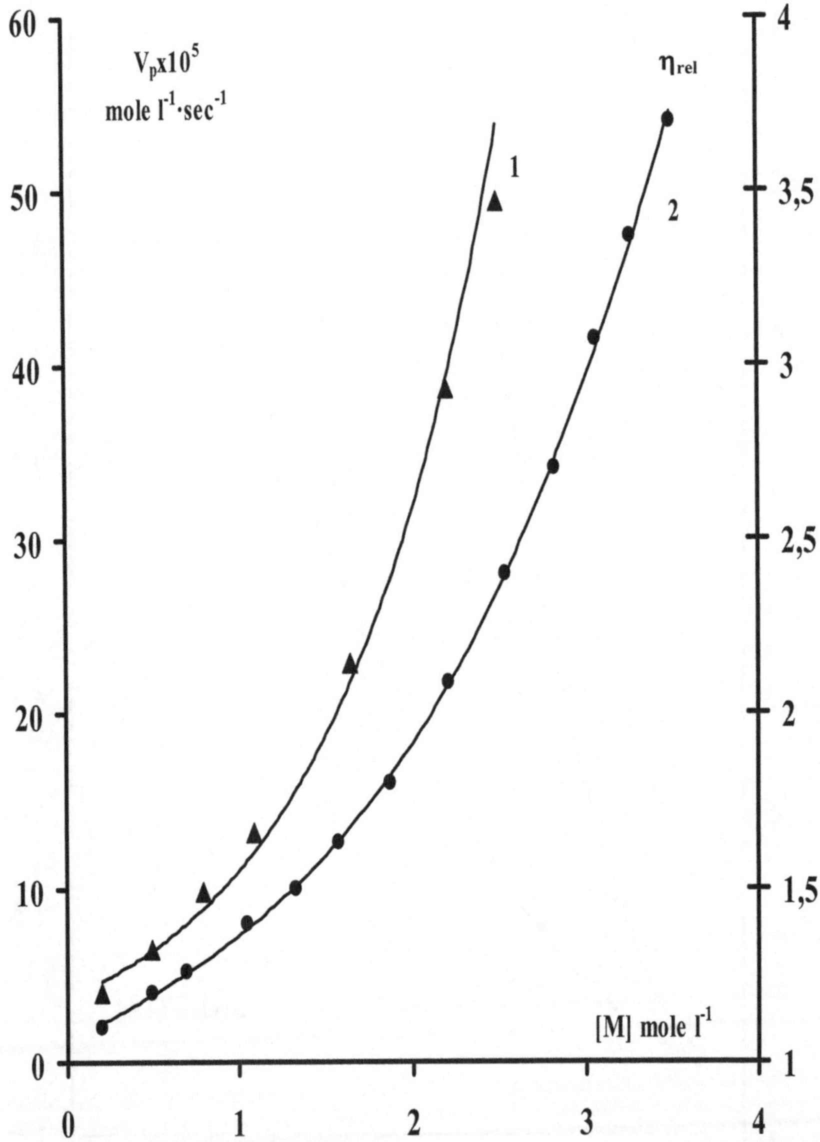

Figure 2.5. The dependence of MAG polymerization initial rate (curve **1**) ([APS] = 5×10^{-3} mole l^{-1}; 60°C) and relative viscosity of initial MAG solutions (curve **2**) (1% hydroquinone; 30°C) on monomer concentration in water solutions.

It is obvious from Figures 2.4 and 2.5 that for mentioned monomer salts GA and MAG the non-linear increase of relative viscosity ($\eta_{rel.}$) is naturally observed with the rise of monomer concentration in initial reaction solution. (As it was already mentioned the solution remained transparent.) Thus, there is definite correlation in the character of polymerization initial rate change and values of $\eta_{rel.}$.

The obtained data (namely the non-linearity of change of $\mathbf{V_p}$ on [**M**]) may be explained by the fact that due to comparatively high values of $\eta_{rel.}$ of initial solutions of monomer salts (at [**M**] > 1 mole·l^{-1}) the constant of chains bimolecular termination rate $\mathbf{k_t}$ even at conditions of very small conversions is turned to be sensitive to the viscosity of initial reaction solutions and consequently to monomer concentration. Mentioned change of values of relative viscosity with the rise of initial monomer concentration should lead to cymbatic reduction of constant $\mathbf{k_t}$ and thus to mentioned non-linear increase of polymerization initial rate. The suggested explanation may be checked experimentally. If we assume that in the studied system the constant $\mathbf{k_t}$ is naturally depended on monomer solution viscosity then in accordance with North's conceptions we should take that $\mathbf{k_o} \sim \eta^{-1}$. Then polymerization rate may be calculated by equation 1 (Section 2.1.1).

The experimental data of dependence of GA (MAG) initial polymerization rate on initial monomer concentration were processed then in coordinates of equation 1.

It turned out that these non-linear dependences are straighten (Figures 2.6 and 2.7) in coordinates lg**V** / lg{[**M**] $\eta_{rel}^{1/2}$ }. The fact of good straighten of experimental curves in coordinates of given equation tells us about correctness of assuming that under determination of reaction order by monomer in given systems it is necessary to consider the viscosity factor.

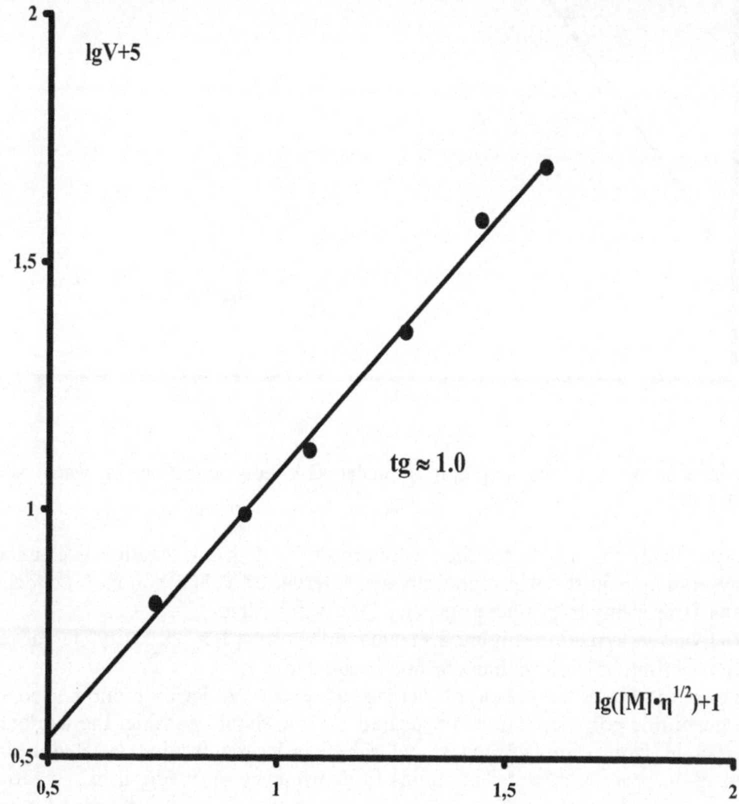

Figure 2.6. The dependence of lgV on lg{[M]·$\eta^{1/2}$} under MAG polymerization in water solutions, [APS] = 5·10^{-3} mole/l; 60°C.

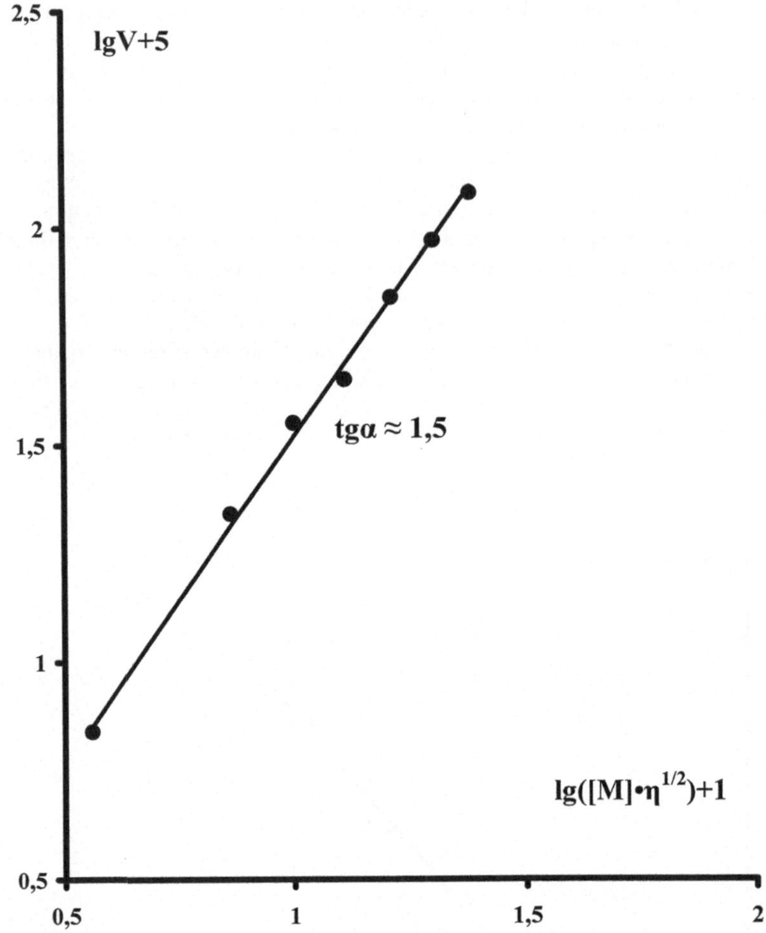

Figure 2.7. The dependence of lg**V** on lg{[**M**]·**η**$^{1/2}$} under GA polymerization in water solutions, [APS] = $5\cdot10^{-3}$ mole/l; 60°C.

Then obviously considering this factor the studied reaction of polymerization is characterized by the first order by monomer in the whole investigated interval of MAG monomer concentration (Figure 2.6), in spite of observing micro-heterogeneity for given system.

In the case of GA polymerization (Figure 2.7) even with correct for "viscosity" factor reaction order by monomer concentration is more than one and is equal to ≈ 1,5.

Naturally the question about the reason of increase of reaction order by monomer concentration under acrylate-guanidine polymerization is appeared. So, we should consider the works by Kabanov and Topchiev et. al. [for example 1 and 3] and see the order regularities revealed during AA and MAA polymerization process in water mediums in dependence on medium pH. It turned out that under MAA polymerization the "classical" order regularities – the half order by initiator (AIBN) and the first by monomer are remained (see also Section 2.2.1). In the case of AA polymerization the half order by initiator and 1,5 order by monomer are observed in the whole studied interval of pH (as in acid, so in alkaline medium). These parameters are maintained independently on neutralizing agent, i.e. in those cases when medium pH was set not only by NaOH addition, but also by TEA and ammonium hydroxide. Earlier Medvedev et. al. [99] also observed sesquialteral order of reaction rate by AA concentration under its polymerization in water solutions.

Some other works also indicated on "anomalous" orders of reactions by concentration of some monomers. So, under polymerization of salt of vinyl pyridine (2VP) and nitrogen acid in H_2O the reaction order by monomer is close to 3, and for salt 2VP and HCl it is close to 4,5 [100]. Loginova et. al. established [101] that in dependence on concentration of monomer (as far as it was diluted) diethylaminoethylmethacrylatehydrochloride the reaction order is changed from 4,7 at $2{,}5 < [\mathbf{M}] < 3{,}7$ mole l^{-1} to 1,2 at $0{,}94 < [\mathbf{M}] < 1{,}6$ mole l^{-1}. At that the decrease of molecular masses of resulted polymers is observed for given system with the rise of monomer concentration. In presented works the authors couldn't explain the observed results.

When studying kinetics of radical polymerization of non-ionogenic monomer N-vinylpyrrolidone (N-VP) in water solutions a lot of authors noticed extreme character of dependence of $\mathbf{V_p}$ on monomer concentration [102–108]. At that the first order by monomer is observed only in diluted monomer solutions $[VP] \leq 1$ mole l^{-1}. In the interval of concentrations $1 \leq [VP] \leq 5 - 6$ mole l^{-1} the variable (less than one) order is observed and in more concentrated solutions $[M] \geq 6 - 9{,}4$ mole l^{-1} the "negative" order by monomer is noticed. (The last number is related to pure VP, i.e. polymerization was carried out in mass).

Authors of recently published work [109] succeeded in establishing of reason of "non-triviality" of mentioned polymerization system in water mediums. It turned out that values of specific contraction under polymerization of VP in H_2O were not constant and changed with the change of monomer concentration in the initial reaction solution that was the consequence of formation of various in structure and strength VP associates with water molecules and which led to deviation from linear dependences of relative densities of both monomer and polymer with the change of VP concentration in water. On the base of said above it was obvious that the use of one fixed value of contraction under calculation of polymerization rates at various VP concentrations (measured by dilatometer method) led to incorrect calculations of $\mathbf{V_p}$ on [VP] that gave anomalous results under determination of reaction order by monomer VP.

A lot of works (for example [110]) are devoted to establishing of possible reasons of deviation of kinetic order regularities from "classic" standards in accordance with equation 2.

$$\mathbf{V_p}=\mathbf{k_p}\times\mathbf{k_t}^{-1/2}\times[\mathbf{M}]\times\mathbf{V_{in}}^{1/2} \qquad (2)$$

Very often at that the systems are considered for which in polymerization processes the separation of phase is observed. In such cases the general laws of kinetics applicable to homogeneous systems are not always carrying out. As it was showed separation of phase was accompanied by the increase of reaction rate often enough that was connected with formation of new phase. In general case on polymerization rate accompanying by separation of new phase will influence a lot of factors and this factor is described in details in Gladyshev's et. al. work [111, 112]. First of all we shall consider what differences are characteristic for heterogeneous polymerization in comparison with homogeneous. As a rule they are the following: the increase of reaction order by initiator from 0,5 up to 1,0; the sharp rise of polymerization rate with the rise of conversion (auto-catalytic process); the decrease of termination rate constant and the increase of molecular masses of resulted polymers. The increase of reaction order by initiator from 0,5 up to 1,0 is observed as under heterogeneous polymerization, so at the stage of high conversions under homogeneous polymerization in the region of so-called "gel-effect" that is connected as a rule with the decrease of $\mathbf{k_t}$ caused by the reduction of probability of meeting of polymer radicals occluded in balls. In the case of systems considered by us under GA and MAG polymerization at small conversions ($\mathbf{q} \leq 5\%$) even at conditions of micro-heterogeneity as it was shown in Figure 2.1 the "classical" order of reaction rate by initiator concentration was remained that testified to bimolecular mechanism of chains termination. The observing for a lot of monomers increase of reaction kinetic order by monomer concentration revealed even at conditions of homogeneity in each concrete case obviously has its own reasons which establishing is not always possible. So, for a number of polymerization systems the increase of order may be caused by the presence of initiation rate dependence on monomer concentration which shouldn't be but is often observed either at high initiator concentrations, or at low concentrations of monomer. Obviously, in this case it is also necessary to choose optimal conditions, i.e. to establish the diapason of initiator concentrations at which the rate of initiation will be constant [113, 114]. Nevertheless, even at correct choice of optimal conditions of polymerization carrying out at practice the order of reaction by monomer very often is between 1,0 and 1,5 and in rare cases as it was mentioned above the higher order was observed.

The analysis of various polymerization systems shows that "non-classic" monomers behavior in reactions of radical polymerization is revealed as a rule in systems when initial mixture monomer–solvent is not ideal, i.e. intermolecular interactions of monomer and (or) polymer molecules between each other and also their interaction with solvent are characteristic for it. It turned out that monomers containing ionized or functional groups undergoing the mediums effects often relate to such systems; and even if such groups are absent, then the interactions of growing radical with solvent are possible as a result of complex formation under the action of hydrogen bonds, etc. that influences on the dependence of $\mathbf{k_p}$ of chain on structure of reaction solution often leading to deviation from equation 2. In such cases we speak about medium effect and that $\mathbf{V_p}$ is non-linear and the dependence $\mathbf{k_p}\,\mathbf{k_t}^{-1/2}$ on reaction solution structure is observed [113, 114].

For polymerization systems considered by us the deviation from single reaction order by monomer for GA as we think is not connected with possible participation of monomer in initiation stage. Since Kabanov and Topchiev in their works established that rate if acrylate-anion initiation doesn't depend on pH of reaction solution, the value of ion force in alkaline region, monomer concentration and also on the nature of neutralizing agent added into polymerization system for change of pH the reaction order by monomer under AA polymerization nevertheless was equal to 1,5 [1]. The main contribution into given phenomenon obviously is made by specific influence of reaction medium on reactivity of monomer and propagation radical in polymerization processes.

On the base of said above we suppose that presented formulation of polymerization rate with consideration of viscosity effect (equation 1) allows satisfactorily describing of initial stage of AA polymerization reaction at studied conditions.

2.2.2.2. Conformational state of propagating chains under radical polymerization of acrylate- and methacrylate guanidines

The very important characteristic of physical-chemical behavior of polyelectrolytes in solution is conformational state which depends on a great number of factors including nature of solvents, polymer ionization degree, solution ion force, etc.. Investigation of conformational behavior of macromolecules containing charged fragments in the chain causes significant interest of researches not only from theoretical point of view, but also from positions of possible application of such polymers for solution of medical-biological, ecological and other problems [115].

At the same time it is known that conformational state of macromolecules of polyelectrolytes in many cases significantly influences on their reaction ability in chemical reactions [1] that was specially studied under investigation of reactions of polyetherification [116], reactions of globular balls, in particular on the examples of sharp change of catalytic activity of ferments with the change of their conformation [117–119].

Under polymerization of ionized polymers in the course of polyelectrolytes chains formation the change of macromolecules conformational state may influence on polymerization kinetics especially under polymerization in water solutions. It is well known that addition of low-molecular electrolytes significantly influences on polyelectrolytes macromolecules state in water solutions, including conformational one [1, 120]. At that in solution the increase of concentration of counter-ions naturally occurs and the corresponding low-molecular ions bearing the charge of the same sign as the charge of poly-ion are appeared.

Since considered monomers are related to the class of ionogenic ones in such system we may expect the increase of probability of chain propagation act realization consisting in the interaction of likely charged molecules of radical and monomer leading to increase of chain $\mathbf{k_p}$.

In connection with it we studied the influence of ion force of reaction solution on MAG polymerization kinetics. With this aim by the method of dilatometry we measured the MAG polymerization rates in water solutions in the presence of low-molecular electrolyte NaCl (and also in methanol in the presence of LiCl) while varying their concentrations from 0,1 up to 5,0 mole·l^{-1}.

As it is obvious from data presented in Table 2.5 the rise of ion force of reaction solution leads to the increase of MAG polymerization rate. These data are in good agree with conceptions formulated above about specific particularities of influence of ion force of solution on polymerization kinetics in the framework of formulated by authors of [1] hypothesis.

Table 2.5. The influence of low-molecular electrolyte NaCl additives on the values of MAG polymerization rate and intrinsic viscosity of PMAG samples resulted from MAG radical polymerization ([APS] = 5,0·10^{-3} mole l^{-1}; H_2O; 60°C)

No.	In the absence of NaCl			In the presence NaCl			
	[M], mole l^{-1}	$V_p \times 10^4$, mole l^{-1} sec^{-1}	$[\eta]^1$, dl g^{-1}	[M], mole l^{-1}	[NaCl], mole l^{-1}	$V_p \times 10^4$ mole l^{-1} sec^{-1}	$[\eta]^{1,2}$, dl g^{-1}
1	0,25	0,4	0,07	-	-	-	-
2	0,50	6,0	0,28	0,50	0,50	-	0,33
3	0,75	1,0	0,32	-	-	-	-
4	1,00	1,1	0,50	1,00	0,50	1,4	0,63
5	1,50	2,2	0,66	-	-	-	-
6	2,00	3,8	1,35	2,00	4,00	4,8	2,00
7	2,50	5,1	1,90	-	-	-	-

Notes: [1] – the viscosity was measured in 1n solution of NaCl at 30°C; [2] – the viscosity of samples obtained in the presence of NaCl.

It is important to note that molecular masses of resulted polymers were also increased with the rise of ion force of reaction solution and we could estimate it by the values of intrinsic viscosity of resulted at that polymer PMAG.

Thus, the influence of ion force of reaction solution on polymerization process of NAG is obvious. However it turned out that additions of salts were not inert since in their presence in MAG molecule the reaction of guanidine ion change by low-molecular cation partially proceeds (that was confirmed later by element analysis). Finally, during the MAG polymerization process in the presence of NaCl in water mediums (or in the presence of LiCl in methanol) the resulted polymer product was not homopolymer – PMAG and it was practically copolymer containing unit of both methacrylate-guanidine and methacrylate sodium (lithium).

For revealing of influence of conformational state of macro-radicals on kinetics of radical polymerization of acrylate- and methacrylate-guanidines in water mediums with the help of viscosimetry method the values of macroscopic viscosities in solutions modeling reaction mixtures at low conversion degrees were measured and obtained data were compared with kinetic ones.

By measuring of reduced viscosity $\eta_{red} = \eta_{sp} / c$ in dependence on monomer concentration in model systems we showed that even at small additions into investigated polymers PAG and PMAG water solutions of corresponding to them monomer salt the sharp change of linear sizes of polymers chains, i.e. their "compression" was observed, and at high concentrations the solution became opaque (Figures 2.8 and 2.9). We also established that in the case of PMAG the opalescence of model solution was observed in monomer solutions diluted enough at concentration [MAG] > 0,4 mole·l^{-1} (Figure 2.9); for the PAG insignificant haze of solution was observed at essentially bigger concentrations of monomer [GA] > 1,3 mole·l^{-1} (Figure 2.8). We should note that close to these concentration regions (regions of transition from total homogeneity to micro-heterogeneity), according to viscosimetry data, the maximum compression of balls of PAG and PMAG macromolecules occurs and the lowest value of η_{red} corresponds to this fact. At further increase of monomer concentrations in the system PAG + GA the solution becomes opaque (turbid) but separation of phase at that was not observed even at significant excess (~ 16 times) of GA monomer concentration ion relation to polymer. In given region with the rise of ion force of model solution together with the appearance of micro-heterogeneity the increase of η_{red} is obviously occurs that may be the consequence of two phenomena – either of transition of system to micro-heterogeneity, or of the reduction of PAG chains sizes as a result of specific complex formation between monomer and polymer molecules (in particular under interaction of functional and guanidine groups of monomer and polymer molecules, Figure 2.10).

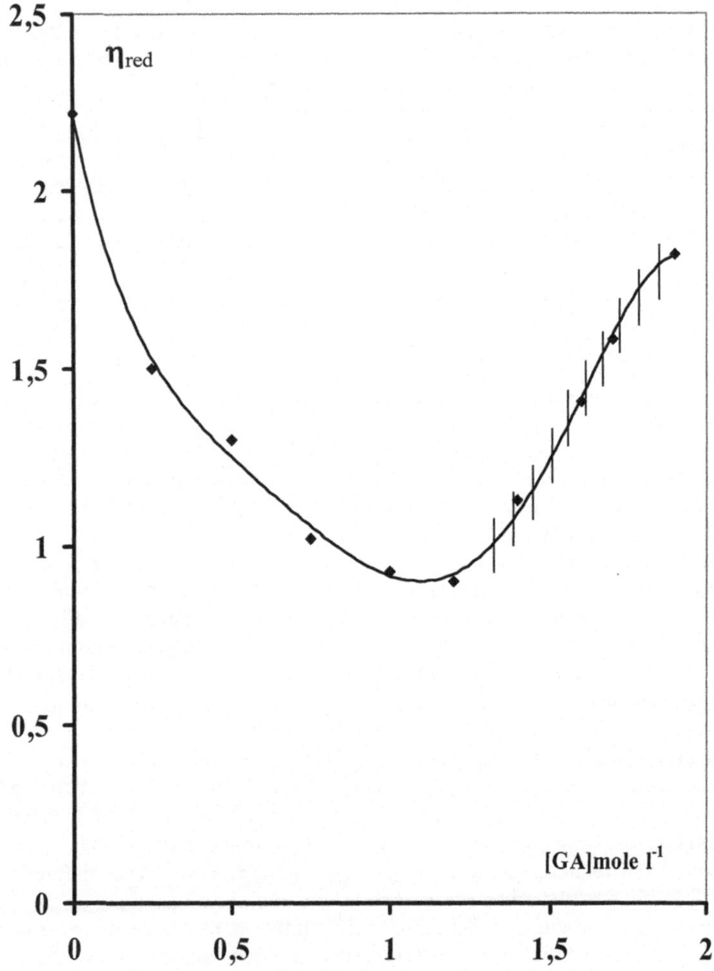

Figure 2.8. The dependence of reduced viscosity of PAG on GA concentration in water at 30°C; $[\eta]_{PAG}$ = 0,27 dl g^{-1}; [PAG] = 0,12 mole l^{-1} (1,62% solution), ///// – the region of heterogeneity.

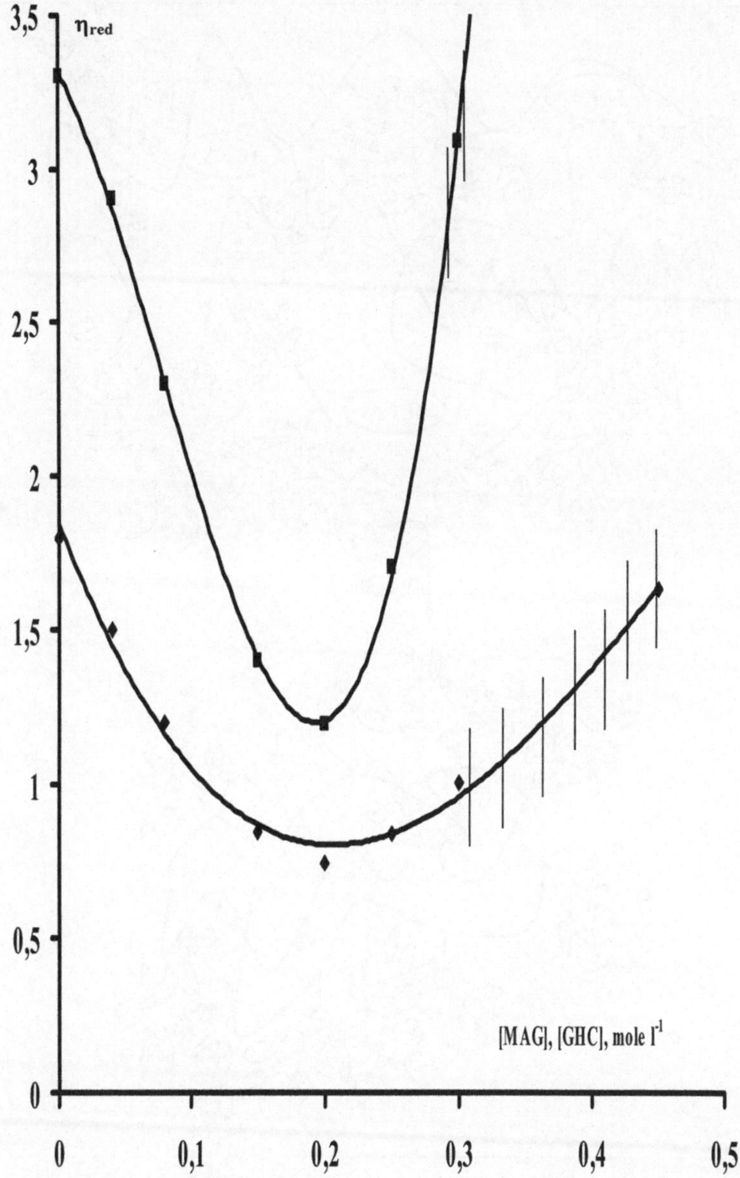

Figure 2.9. The dependence of reduced viscosity of PMAG model solution in systems: **1** – PMAG + MAG; **2** – PMAG + GHC ($[\eta]_{PMAG}$ = 0,27 dl g^{-1}; 1,7% solution of PMAG in water; 0,1 mole·l^{-1}; 30°C, hydroquinone), ///// – the region of heterogeneity.

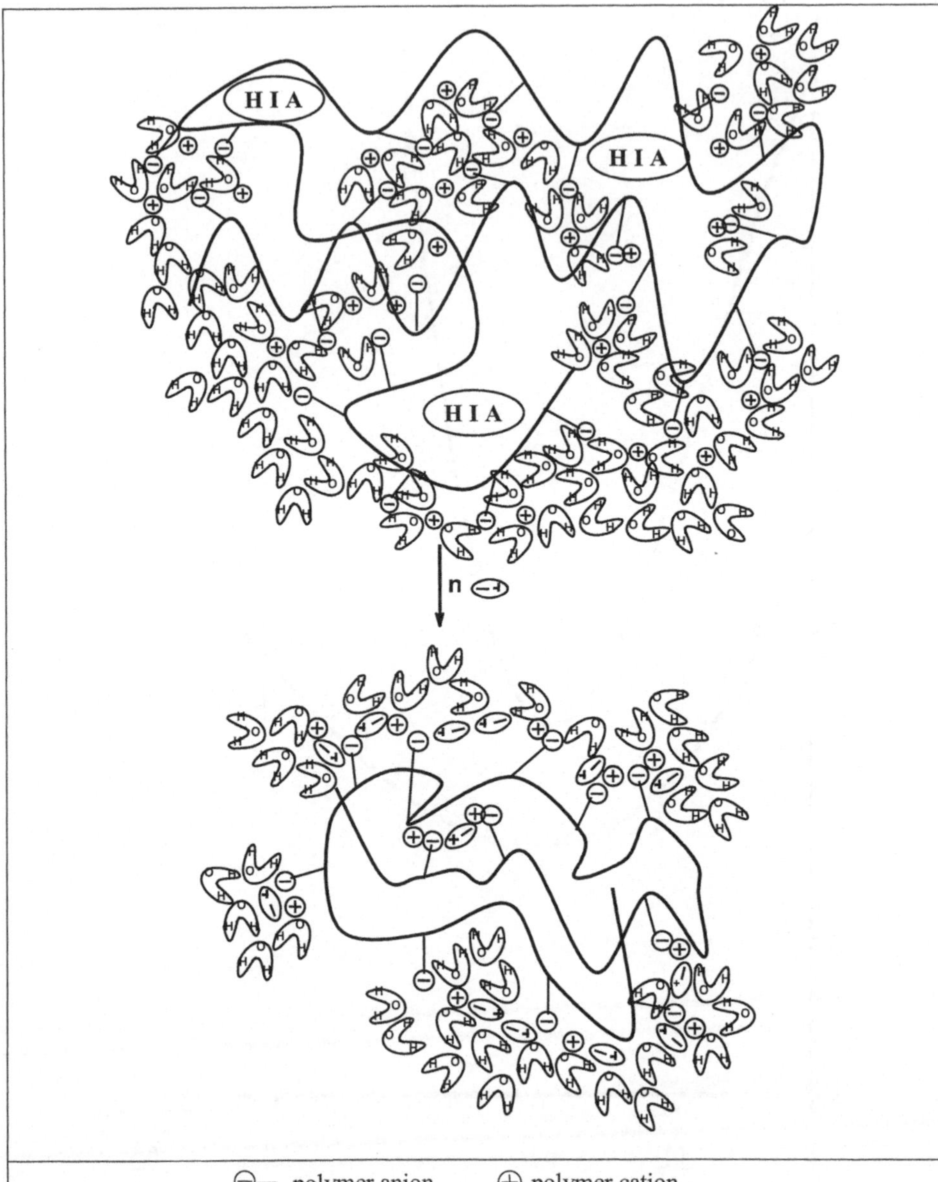

Figure 2.10. The change of PAG and PMAG balls' form under the action of monomer or guanidinehydrochloride (model system).

In the other model system PMAG + MAG at the transition of η_{red} through the minimum value ([MAG] > 0,4 mole·l^{-1}) complete turning of PMAG chain with phase separation is observed (precipitation of white residue is clearly observed visually) that is revealed at relatively low (thrice-repeated) excess of MAG monomer in relation to PMAG. Since due to PMAG sediment precipitation the further carrying out of viscosimetry investigations becomes impossible (viscosimeter capillary is clogged) the further measurements were stopped at relatively low concentrations of monomer. Nevertheless, the analysis of data received by viscosimeter investigations allows concluding that there was a good coincidence with kinetic data.

Actually, in both cases the interval of monomers concentrations in which reaction systems keep their homogeneity is practically coincides (of course with definite correction by temperature course of investigated processes since polymerization kinetics as it was mentioned earlier was measured at 60°C, and viscosimetry studies were carried out at 30°C). Moreover, undoubtedly the correspondence of "transition region" of monomer concentrations values was also traced at which compression of macroradicals balls and their turning obtained by mentioned independent methods (dilatometer and viscosimetry) was observed.

For revealing of possible reasons explaining so uncharacteristic for ionogenic polymers in water solutions "salting-out" at low degrees of transformation under the action of proper monomer we carried out kinetic investigations of GA and MAG polymerization from initial up to high conversions (for a number of samples up to 80%) by dilatometry method. At that the solution's transparency at investigated conditions was fixed visually. We should note that for some monomer concentrations at high conversion degrees as it is known due to high viscosity of reaction solution such measurements by dilatometry method are impossible. In other cases when viscosity of reaction solution is significantly increased with conversion but solution keeps fluidity such measurements are possible however they reflect mainly the qualitative picture of proceeding processes.

As it was shown, under GA polymerization (0,25 < [GA] ≤ 1,3 mole·l^{-1}) kinetic dependences of $\mathbf{V_p}$ on **q** % were linear, and polymerization solution remained completely transparent up to high conversions. At higher monomer concentration [GA] > 1,5 mole·l^{-1} the linearity of given dependence is observed only at relatively low conversions (**q** ≤ 15%). At the same time from the initial conversions the micro-heterogeneity is appeared. The fact that as far as the monomer is spent in reaction solution at high conversion degrees solution turbidity is gradually reduced and solution becomes absolutely transparent is amazing.

The given phenomenon observed under GA polymerization in water is completely uncharacteristic for MAG. The homogeneity in the case of MAG was fixed only in diluted solutions [M] ≤ 0,4 mole·l^{-1} (but remained up to high conversion degrees). At high initial MAG concentrations at the initial section of kinetic curve the micro-heterogeneity was observed (at $\mathbf{V_p} \approx$ const, **q** ≤ 5%). At **q** ≥ 20% the PMAG sediment is precipitated in dilatometer and complete heterogeneity remains up to deep conversions (approximately for 10 hours of polymerization up to complete monomer spending in reaction system) (for mentioned systems linearity of kinetic curves remains at **q** ≤ 20%).

The analysis of results described above shows that MAG behavior in polymerization reaction has the features differed from GA behavior at the same conditions. The main feature of these differences as we assume is in special structure of PMAG chains in water mediums. For understanding of reasons of such "non-ordinariness" of MAG behavior we probably should compare particularities of PAA and PMAA conformational behavior in water solutions. In fifties they established that PMAA behavior in water solutions was slightly differed from classic behavior of other synthetic polyelectrolytes [121–124].

Then under investigation of titration curves of PAA and PMAA they showed that the anomalous course of curves of potentiometric titration of PMAA [125–127] (their clear non-linearity) was observed and the dependence of PMAA solutions viscosity at low degrees of ionization on ionization value $\boldsymbol{\alpha}$ was absent [81]. This fact allows concluding that non-charged macromolecule of PMAA in water solutions is characterized by coagulated average conformation which then during the ionization process is opened in comparatively narrow region of dissociation degrees α ($0,1 < \alpha < 0,15$), i.e. PMAA molecules undergo cooperative conformational transition. The given phenomenon as we suppose is connected with destruction of secondary structure of PMAA molecules in water solutions at low values of α. Structured state of PMAA macromolecules is caused as we established by non-polar interactions between methyl groups.

Later Nekrasova et. al. [128] while considering PMAA chains models showed that the structure with slip plane when all methyl groups were situated on the one side and carboxyl ones on the other side transversely to slip plane passing through molecule axis corresponded to the highest interaction of methyl group of given macromolecules.

Theoretical calculations [129] show that the minimum of potential energy corresponds to such structure and thus on the base of analysis of numerous experimental and theoretical data they propose the hypothesis that structured state of PMAA macromolecules in water solutions is determined by hydrophobic interactions. In other solvents hydrophobic interactions are absent and that is why a lot of anomalies observing in PMAA water solutions are not found in organic solvents at all. Moreover, under addition of methanol, dioxane, DMFA into water solutions of non-ionized PMAA the sharp increase of $[\eta]_{intr}$ in the narrow interval of solvent structure change occurs that testifies to conformational transition. This transition is accompanied by both the increase of macromolecules linear sizes and by the change of closest order [130, 131]. Thus, addition of organic solvents as well as ionization reduces stability of structured state of PMAA macromolecule in water solutions. For PAA macromolecules such cooperative conformational transition in water solution is absent. At the same time at essentially high ion force of PAA or PMAA solutions or under addition of low-molecular salt into polymers solution the reduction of linear sizes of polymer chains occurs and they become ball-shaped [132]. Such detailed discussion of particularities of PMAA and PAA conformational behavior is seemed for us necessary for deeper understanding of particularities of behavior in monomers water solutions of the same type as AA and MAA but with guanidine fragment in the form of counter-ion.

Lets consider those particularities which differ guanidine containing poly(meth)acrylates from PAA and PMAA (Figure 2.11). As we established by IR-spectroscopy method when comparing spectra of sodium salt of methacrylic acid and MAA in the case of MAA high delocalization of electrons by $C\text{–}O^-$ bond in methacrylate-ion was observed (we also showed, that natures of ion connection in monomer and polymer were equal).

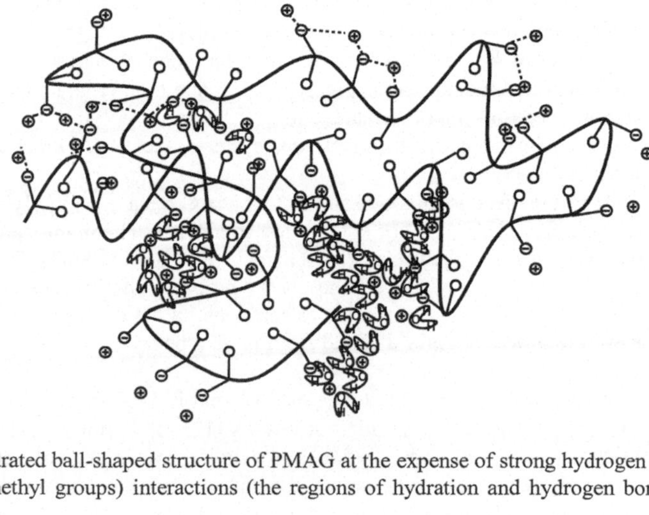

(**a**) Compressed hydrated ball-shaped structure of PMAG at the expense of strong hydrogen bonding and hydrophobic (O— methyl groups) interactions (the regions of hydration and hydrogen bonding are shown fragmentary).

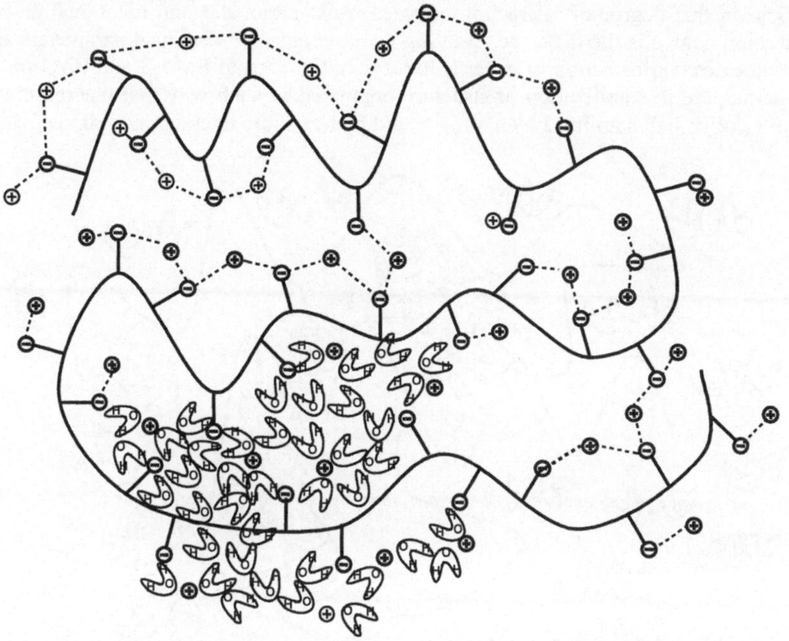

(**b**) Hydrated flocculent ball-shaped structure of PAG (the regions of hydration and bonding are shown also fragmentary).

Figure 2.11. The structures of PMAG (**a**) and PAG (**b**) homopolymers (---- – hydrogen bonds, the rest legend as in Figure 2.10).

Thus, PMAG and PAG due to such delocalization may be considered as polymers in which counter-ion is strongly bonded with carboxyl group. From the other hand, guanidine groups are the important fragments of a number of nature compounds, including amino-acids – arginine, folic acid, numerous proteins and nucleic acid. And such singularity of conformational behavior of synthesized PAG and PMAG polymers in water solutions as we think allows assuming definite analogue of conformational transformations of considered by us macromolecules and proteins systems. The established earlier in a lot of works fact about denaturation of a number of nature proteins and also of some synthetic polymers – analogues of proteins (for example, poly-N-vynilpyrrolidone – PVP) [120, 133–135] under the addition of low-molecular guanidine salts into their solutions serves as the additional argument of the possibility of such analogue. Authors propose that observing sharp turning of PVP chain is possible due to the presence of two amino-groups in guanidine molecule able to compete with carbonyl group of polyvynilpyrrolidone blocking its further interaction with solvent molecules (water).

So, in the presence of guanidine hydrochloride the intrinsic viscosity of PVP in alcohol solutions is markedly reduced. The value of $\mathbf{k}[\boldsymbol{\eta}]^2$, i.e. the magnitude characterizing the interaction between polymer and solvent molecules is changed especially sharply; at that the PVP molecules practically completely soluble in alcohol become insoluble in the presence of guanidine hydrochloride that is the consequence of blocking of pyrrolidone cycle oxygen by guanidine hydrochloride molecules leading to the increase of forces of inter-molecular association of PVP cycles at the expense of hydrophobic interactions. Moravetz and other authors [120, 136] studied the influence of various factors on protein denaturation in details established that various guanidine salts and also the urea exert strongly denaturation influence on proteins molecules under their introduction into the solution even in small amount ~ 1%.

It is known that degree of interaction between water molecules and functional groups of PE, i.e. PE hydration degree is the factor determining its basic physical-chemical parameters and first of all macromolecules conformation in water solutions. In the case of PMAG and PAG molecules of H_2O also participate in stabilization of structure organized in such way that macromolecule takes the shape of swollen ball stabilized by hydrogen and hydrophobic interactions (Figure 2.12).

Figure 2.12. Change of conformation and form of homopolymer (PAG and PMAG) under the action of mutual monomer in the course of polymerization as a result of specific interactions between polymer, monomer and water molecules (legend as in Figure 2.10).

Introduction of monomer containing guanidine groups leads to "disorientation" of mentioned structure at the expense of primary interaction of NH_2-groups with carboxyl group as a result of which the water molecules are removed from inter-chain interactions, functional carbon chain groups are "blocked" and finally the macromolecule is compressed becoming more compact in the shape of coagulated conformation of globular type. The intrinsic viscosity of polymer solution at that is reduced: in the case of PAG the solution becomes opaque (turbid) and in the case of PMAG at high conversions **q**% or at high monomer concentrations the separate phase is formed.

Thus, we assume that the observing in the radical polymerization of MAG micro-heterogeneity is the consequence of the fact that MAG monomer at the expense of guanidine groups content leads to coagulation of PMAG propagating chains as a result of cooperative conformational transition.

Of course if we shall draw an analogy close to mechanism of guanidine salts action on macromolecules of synthetic (or nature) proteins and considered polymer systems then the PMAG to a definite degree may be considered as such analogue. Since stability of proteins and proteins-similar synthetic polymers structures is provided mainly not by hydrogen bonds (their role is undoubtedly in the creation of proteins structure), but by hydrophobic interactions. That is why as we assume the presence of MAG monomer "excess" in polymerization process leads to the change of "ordered" PMAG structure which is formed in water solutions in the absence of monomer or at its low content (less than 0,4 $mole \cdot l^{-1}$) (Figure 2.12).

In the case of PAG the total destruction of macromolecule structure is not observed and it is more labile since as we discussed earlier even at significant excess of GA monomer in PAG solution the system remained homogeneous (sediment was not formed) and at the decrease of GA monomer concentration in reaction solution (during its consumption in polymerization process) the polymerization solution gradually became transparent again and "micro-salting-out" of macromolecules was stopped.

The analysis of data obtained by us allows concluding that observing changes of macromolecules conformations (at any case for PAG) are determined mainly by the concentration of monomer salt in reaction solution, and weakly depend on molecular mass of corresponding polymer (that we may expect a priory). Otherwise it is difficult to explain the observing in polymerization process phenomenon of transition of reaction system PAG + GA (at high GA monomer concentrations and higher **q**%) from completely opaque to totally transparent. It occurs as monomer salt is spent in reaction solution during polymerization process against increasing η_{rel}, i.e. returning of PAG + GA system (and obviously PAG conformation) to initial homogeneous characteristic as it was mentioned for low [GA] and small **q**%.

It is interesting to estimate also possible influence on conformation of PMAG in water solution of its low-molecular analogue. With this aim we select guanidine hydrochloride (GHC). As it is obvious (Figure 2.9 curve 2) the dependence of reduced viscosity on salt concentration in system PMAG + GHC and dependence in model system PMAG + MAG (Figure 2.9 curve 1) are symbate. In both cases under addition of salt the coagulation of PMAG chains in a narrow interval of solution's ion force change (~ 0,10–0,15) is observed that is caused probably by phase transition of system as a result of disorientation of macromolecules structure, by the way this process is irreversible.

It is interesting that in the system PMAG + GHC at [GHC] > 0,3 $mole \cdot l^{-1}$ the gelatinous opaque mass is formed that is caused probably by the formation of spatial net structure.

Thus, the characteristic for investigated polymer systems conformational transformations accompanying by reconstruction of chain local structure are caused as it is obvious by "blocking" action of guanidine monomer groups leading to essential coagulation of PMAG macromolecules, and particularities of conformational behavior of PMAG in contrast to PAG are explained by significant contribution of hydrophobic interaction in PMAG chains that impacts to given macromolecules order in structure organization and lower lability in comparison with PAG.

It is remarkable that in the case if we shall level "denaturating" action of guanidine group as we did under copolymerization of GA and MAG with DADMAC in water mediums then copolymerization reaction (for all structures of reaction solution in the case of GA and up to 70% of acrylic comonomer content in the case of MAG) up to high conversion degrees (~ 60%) proceeded at homogeneous conditions.

In other words as in the case of nature protein molecules the introduction of units of "strange" "neutral" monomer into copolymer (in our cases it may be MAA, sodium methacrylate or DADMAC) leads to breaking of polymer chain tacticity. The higher the number of such "inclusions" into PMAG chain, the lower the influence of guanidine containing monomer on heterogeneity of MAG polymerization process.

The same picture – the heterogeneity in polymerization process – was observed also under MAA and AA polymerization in water solutions under the addition of amines ethylenediamine, isobutylamine and pyridine as neutralizing agents into reaction solution for setting of definite pH [3, 73, 74]. Authors explained this phenomenon by specific complex formation between MAA monomer molecules and corresponding amine and by probable contribution of hydrophobic interactions which additionally stabilized ion pairs at the ends of ionized radicals of PMAA propagation.

In our case we should say that the reason of observing in investigated systems microheterogeneity is "denaturating" action of guanidine groups of mutual monomer at their definite "excess" which are able to destruct the forming hydrated structure of macromolecules formed during polymerization process that leads to significant conformational changes accompanying in a number of cases by the lost of solubility and "salting-out" of polymer.

References

1. V.A. Kabanov, D.A. Topchiev, *polymerization of ionized monomers*, Moscow: Nauka (1975) (*in Russian*).

2. D.A. Topchiev, Yu.A. Malkanduev, *Radical polymerization of N,N-dialkyldiallyl ammonium of halogenids*, Nal'chik: KBGU (1997) (*in Russian*).

3. D.A. Topchiev, Dissertation of doctor of chemical sciences, Moscow: INKhS (1973) (*in Russian*).

4. V.A. Kabanov, V.P. Zubov, Yu.D. Semchikov, *Complex-radical polymerization*, Moscow: Nauka (1987) (*in Russian*).

5. G.B. Butler, *J. Polym. Sci.*, **48**, No.1, 279 (1960).

6. M.D. Burtnett, G.B. Butler, *J. Org. Chem.*, **25**, 309 (1960).

7. G.B. Butler, *Cyclopolymerization and Cyclocopolymerization,* New York: Marsel Dekker (1992).

8. G.C. Corfield, *Chem. Soc. Rev.*, **1**, No. 3, 523 (1972).

9. G.B. Butler, *Amer. Chem. Soc. Div., Polym. Chem. Preprints.*, **8**, No.1, 35 (1967).

10. G.B. Butler, S. Kimura, *J. Macromol. Sci. Chem.*, **5A**, No.1, 181 (1971).

11. G.B. Butler, A. Crawshow, W.L. Miller, *J. Am. Chem. Soc.*, **80**, No.14, 3165 (1958).

12. M. Julia, M. Maumy, *Bull. Soc. Chim. Fr.*, **1**, 434 (1966).

13. M. Julia, *Chem. Eng. News.*, **41**, 100 (1966).

14. G.B. Butler, *J. Am. Chem. Soc.*, **89**, 35 (1967).

15. H.G. Richey, A.M. Rothman, *Tetrahedron Lett.*, **12**, 1457 (1968).

16. N.O. Brace, *J. Polym. Sci.*, A-1, **8**, No. 8, 2091 (1970).

17. J.E. Lancaster, L. Baccei, H.P. Panzer, *J. Polym. Sci., Polym. Lett. Ed.*, **14**, No. 9, 549 (1976).

18. W.E. Gibbs, J.M. Barton, *in Kinetics and Mechanism of Polymerization*, Ed. Hat G.E. New York: Dekker, part 1, chapter 2 (1978).

19. H.L. Panzik, J.E. Mulvaney, *J. Polym. Sci., Polym. Chem. Ed.*, **10**, No. 12, 3469 (1972).

20. K. Uzushido, A. Matsumoto, M. Giwa, *J. Polym. Sci., Polym. Chem. Ed.*, **16**, No. 5, 1081 (1978).

21. T.F. Gray, G.B. Butler, *J. Macromol. Sci. Chem.*, **9A**, No. 1, 45 (1975).

22. A. Matsumoto, J. Tamura, M. Jamawak, M. Oiwa, *J. Polym. Sci., Polym. Chem. Ed.*, **17**, No. 5, 1419 (1979).

23. S.R. Johns, R.I. Willing, S. Middleton, A.K. Ong, *J. Macromol. Sci. Chem.*, **10A**, No. 5, 875 (1979).

24. R.M. Ottenbreit, *Ing. Engng. Chem. Prod. Res. Dev.*, **19**, 520 (1980).

25. L.M. Bouman, C.I. Cha, *J. Polym. Sci. Polym. Lett. Ed.*, **17**, No. 3, 167 (1979).

26. C. Wandrey, *Acta Polym.*, **32**, 177 (1981).

27. D.A. Topchiev, G.T. Nazhmetdinova, A.M. Krapivin, V.A. Shreider, V.A. Kabanov, *Vysokomol. Soed.*, **24B**, No. 6, 473 (1982) (*in Russian*).

28. D.H. Solomon, *J. Macromol. Sci. Chem.*, **9A**, No. 1, 97 (1975).

29. D.G. Hawthorne, S.R. Johns, R.I. Willing, *Aust. J. Chem.*, **29**, No. 9, 315 (1976).

30. S.R. Johns, R.I. Willing, S. Middleton, A.K. Ong, *J. Macromol. Sci. Chem.*, **10A**, No. 5, 875 (1979).

31. D.G. Hawthorne, S.R. Johns, D.H. Solomon, R.I. Willing, *Aust. J. Chem.*, **3**, No. 215, 1155 (1979).

32. A.L. Beckwith, A.K. Ong, D.H. Solomon, *J. Macromol. Sci. Chem.*, **A**, No. 9, 125 (1975).

33. A.L. Beckwith, D.G. Hawthorne, D.H. Solomon, *Aust. J. Chem.*, **29**, No. 9, 995 (1976).

34. D.H. Solomon, *J. Polym. Sci. Polym. Symposium.*, **49**, 175 (1975).

35. S.D. Haman, A. Pompe, D.H. Solomon, T.H. Spurling, *Aust. J. Chem.*, **29**, No. 9, 1975 (1976).

36. G. Moad, D.H. Solomon, *Chemistry of free radical polymerization.*, Oxford: Pergamon (1995).

37. D.A. Topchiev, G.T. Bikasheva, A.I. Martynenko, N.M. Kaptzov, L.A. Gudkova, V.A. Kabanov, *Vysokomol. Soed.*, **22B**, No. 4, 269 (1980) (*in Russian*).

38. D.A. Topchiev, G.T. Bikasheva, A.I. Martynenko, N.M. Kaptzov, L.A. Gudkova, V.A. Kabanov, *Polymer amines: monomers synthesis, polymerization and ways of application national economy*, Moscow: Nauka (1980) (*in Russian*).

39. G.T. Nazhmetdinova, V.A. Shreider, D.A. Topchiev, V.A. Kabanov, *Izv. AN SSSR, Ser. Khim.*, **5**, 1024 (1984) (*in Russian*).

40. D.A. Topchiev, G.T. Nazhmetdinova, *Vysokomol. Soed.*, **25A**, No.3, 636 (1983) (*in Russian*).

41. D.A. Topchiev, G.T. Nazhmetdinova, V.A. Kabanov, *Izv. AN SSSR, Ser. Khim.*, **9**, 2146 (1989) (*in Russian*).

42. N.A. Babaev, A.I. Martynenko, D.A. Topchiev, V.A. Kabanov, Ch. Wandrey, M. Hahn, W. Jaeger, G. Reinisch, *Acta Polymerica*, **36**, No. 7, 396 (1985).

43. N.A. Golubkova, A.I. Martynenko, N.A. Babaev, A.V. Nechaeva, A.A. Efendiev, D.A. Topchiev, V.A. Kabanov, *Izv. AN SSSR, Ser. Khim.*, **2**, 485 (1986) (*in Russian*).

44. A.I. Martynenko, Ch. Wandrey, W.Jaeger, M. Hahn, D.A. Topchiev, G. Reinisch, V.A. Kabanov, *Acta Potymerica*, **36**, No. 9, 516 (1985).

45. N.A. Golubkova, A.M. Drabkina, L.A. Gudkova, D.A. Topchiev, *Acta Polymerica*, **40**, No.7, 435 (1989).

46. Ch. Wandrey, W. Jaeger, G. Reinisch, *Acta Polymerica*, **32**, No. 4, 197 (1989).

47. D.A. Topchiev, A.I. Martynenko, E.Yu. Kabanov, L.M. Timofeeva, V.D. Oppengeim, A.S. Shashkov, A.M. Drabkina, *Vysokomol. Soed.*, **36A**, No. 8, 1242 (1994) (*in Russian*).

48. V.A. Kabanov, D.A. Topchiev, *Vysokomol. Soed.*, **30A**, No. 4, 675 (1988) (*in Russian*).

49. A.M. North, G.A. Reed, *J. Polym. Sci.*, **1A**, 1311 (1963).

50. R.S. Armentrout, C.L. McCormick, *Polym. Prep.*, **40**, No. 2, 193 (1999).

51. D.B. Thomas, R.S. Armentrout, C.L. McCormick, *Polym. Prep.*, **40**, No. 2, 275 (1999).

52. A. Ali Sk., A. Rasheed, M.M. Wazeer, *J. Polymer.*, **40**, 2439 (1999).

53. D.A. Topchiev, N.A. Sivov, E.J. Goethals, *Russian Chem. Bull.*, **43**, No. 11, 1864 (1994).

54. G.E. Zaikov, Yu.A. Malkanduev, S.Yu. Khashirova, A.M. Esmurziev, A.I. Martynenko, L.I. Sivova, N.A. Sivov, *in Biochemistry and Chemistry: Research and Developments*, New York: Nova Science Publishers, 39 (2003).

55. G.E. Zaikov, Yu.A. Malkanduev, S.Yu. Khashirova, A.M. Esmurziev, A.I. Martynenko, L.I. Sivova, N.A. Sivov, *in Chemical Reactions in Liquid and Solid Phase: Kinetics and Thermodynamics*, New York: Nova Science Publishers, 165 (2003).

56. G.E. Zaikov, Yu.A. Malkanduev, S.Yu. Khashirova, A.M. Esmurziev, A.I. Martynenko, L.I. Sivova, N.A. Sivov, *Russian Polymer News*, **8**, No. 4, 1 (2003).

57. G.E. Zaikov, Yu.A. Malkanduev, S.Yu. Khashirova, A.M. Esmurziev, A.I. Martynenko, L.I. Sivova, N.A. Sivov, *J. Environ. Protect. And Eco*logy, **4**, No. 4, 863 (2003).

58. G.E. Zaikov, Yu.A. Malkanduev, S.Yu. Khashirova, A.M. Esmurziev, A.I. Martynenko, L.I. Sivova, N.A. Sivov, *J.Appl.Pol.Sci.*, **91**, 439 (2004).

59. L.M. Timofeeva, Yu.A. Vasil'eva, N.A. Klesheva, D.A. Topchiev, *Izv. AN SSSR, Ser. Khim.*, No. 5, 865 (1999) (*in Russian*).

60. Yu.A. Vasil'eva, N.A. Klesheva, G.L. Gromova, A.I. Rebrov, M.P. Filatova, E.B. Krut;ko, L.M. Timofeeva,D.A. Topchiev, *Izv. AN SSSR, Ser. Khim.*, No. 3, 430 (2000) (*in Russian*).

61. Yu.A. Vasil'eva, Dissertation of candidate of chemical sciences, Moscow: INKhS (2000) (*in Russian*).

62. M.F. Hoover, *J. Macromol. Sci. Chem.*, **A4**, 1327 (1970).

63. A. Katchalsky, G. Blauer, *Trans Faraday Soc.*, **47**, No.12. p.1360 (1951).

64. A. Katchalsky, N. Shavit, H. Eisenberg, *J. Polymer Sci.*, **13**, 69 (1959).

65. T. Alfrey, C.G.Overberger, S.H. Pinner, *J. Am. Chem. Soc.*, **75**, 4321 (1953).

66. S.H. Pinner, *J. Am. Chem. Soc.*, **74**, No. 2, 438 (1952).

67. G. Blauer, *Trans Faraday Soc.*, **56**, 606 (1960).

68. G. Blauer, *J. Polymer Sci.*, **9**, 167 (1959).

69. G.S. Misra, H. Narain, *Macromol. Chem.*, **113**, 85 (1968).

70. H. Morawetz, I.D. Rubin, *J.Polymer Sci.*, **57**, 669 (1962).

71. V.T. Popov, D.A. Topchiev, V.A. Kabanov, V.A. Kargin, *Vysokomol. Soed.*, **11B**, No. 8, 583 (1969) (*in Russian*).

72. V.A. Kabanov, D.A. Topchiev, *Vysokomol. Soed.*, **13A**, No.6, 23 (1971) (*in Russian*).

73. T.M. Karaputadze, D.A. Topchiev, D.A. Topchiev, *Vysokomol. Soed.*, **13B**, No. 1, 34 (1971) (*in Russian*).

74. D.A. Topchiev, R.Z. Shakirov, L.P. Kalinina, T.M. Karaputadze, V.A. Kabanov, *Vysokomol. Soed.*, **14A**, No. 3, 581 (1972) (*in Russian*).

75. V.G. Popov, D.A. Topchiev, V.A. Kabanov, V.A. Kargin, *Vysokomol. Soed.*, **14B**, No. 11, 117 (1972) (*in Russian*).

76. A.I. Martynenko, Dissertation of candidate of chemical sciences, Moscow: INKhS (1981) (*in Russian*).

77. I.R. Huizenga, P.F. Grieger, F.T. Wall, *J. Am. Chem. Soc.,* **72**, 2636 (1950).

78. A.M. North, *Chemistry and Industry,* 1295 (1968).

79. E.V. Anufriev, M.V. Vol'kenshtain, M.G. Krakovyak, T.V. Sheveleva, *Dokl. AN SSSR*, **182**, 361 (1968) (*in Russian*).

80. T.N. Nekrasova, A.G. Gabrielyan, O.B. Ptitzin, *Vysokomol. Soed.*, **10A**, 297 (1968) (*in Russian*).

81. T.N. Nekrasova, O.B. Ptitzin, M.S. Shikanova, *Vysokomol. Soed.*, **10A**, 1530 (1968) (*in Russian*).

82. T.M. Birshtein, E.V. Anufriev, T.N. Nekrasova, O.B. Ptitzin, T.V. Sheveleva, *Vysokomol. Soed.*, **7**, 372 (1965) (*in Russian*).

83. N.I. Gal'perina, V.F. Gromov, P.M. Khomikovsky, A.D. Abkin, E.N. Zavyalova, *Vysokomol. Soed.*, **16B**, No.4, 287 (1974) (*in Russian*).

84. N.I. Gal'perina, T.A. Tugunaeva, V.F. Gromov, P.M. Khomikovsky, A.D. Abkin, *Vysokomol. Soed.*, **17A**, No.7, 1455 (1975) (*in Russian*).

85. N.I. Gal'perina, V.F. Gromov, V.D. Moiseev, *Vysokomol. Soed.*, **18B**, No.5, 384 (1976) (*in Russian*).

86. V.F. Gromov, P.M. Khomikovsky, A.D. Abkin, *Vysokomol. Soed.*, **12B**, No.10, 767 (1970) (*in Russian*).

87. V.F. Gromov, N.I. Galperina, T.O. Osmanov et. al., *Europ. Polymer J.*, **16**, 529 (1980).

88. G.I. Bayoras, T.N. Alishauskene, N.N. Slavnitzkaya et.al., *Vysokomol. Soed.*, **23B**, No.2, 86 (1981) (*in Russian*).

89. V. Gromov, E.V. Bune, E.N. Teleshov, *Uspekhi khimii*, **63**, No.6, 530 (1994) (*in Russian*).

90. F.S. Shriver, G. de Smets, Van Thielen Y., *J. Polymer Sci.*, **6B**, No.8, 547 (1968).

91. A. Chapiro, R. Gouloubandi, *Europ. Polymer. J.*, **10**, No.12, 1159 (1974).

92. A. Chapiro, D.Goldfoeld-Freidish, J. Perichon, *Europ. polymer J.*, **11**, No.7, 515 (1975).

93. A. Chapiro, J. Dulien, *Europ. Polymer J.*, **13**, No.7, 563 (1977).

94. N.A. Sivov, A.I. Martynenko, E.Yu. Kabanova, N.I. Popova, G.N. Bondarenko, S.Yu. Khashirova, A.M. Esmurziev, *Materials of scientific conference INKhS RAS*, Moscow (2003) (*in Russian*).

95. N.A. Sivov, A.I. Martynenko, E.Yu. Kabanova, N.I. Popova, A.M. Esmurziev, *Materials of the Third all-Russian scientific Kargin's conference "Polymers-2004"*, Moscow: MSU, **1**, 168 (2004) (*in Russian*).

96. N.A. Sivov, A.I. Martynenko, E.Yu. Kabanova, N.I. Popova, S.Yu. Khashirova, A.M. Esmurziev, *Neftekhimiya*, No.1, 47 (2004) (*in Russian*).

97. A.M. Esmurziev, M.P. Filatova, A.N. Sivov, G.N. Bondarenko, M.A. Dzyubina, A.I. Martynenko, N.A. Sivov, *Materials of the all-Russian scientific conference "The structure and dynamics of molecular systems"*, Yal'chik, 343 (2003), 312 (2004) (*in Russian*).

98. N.A. Sivov, A.I. Martynenko, E.Yu. Kabanova, L.I. Sivova, G.E. Zaikov, *J.Appl.Pol.Sci.*, (2005) (*in press*).

99. T.M. Gritzenko, S.S. Medvedev, *Zh. Phyz. Khim.*, **30**, 1238 (1956) (*in Russian*).

100. S. Fujioka, K. Hayashi, S. Okamura, *Annual report of Japan Ass. of Radiation Researsh,* **4**, 199 (1962).

101. N.N. Loginova, R.K. Gavurine, M.L. Alexandrova, *Vysokomol. Soed.*, **11B**, No.19, 643 (1969) (*in Russian*).

102. F.P. Sidel'kovskaya, *The chemistry of N-vinylpyrrolidone and its polymers*, Moscow: Nauka (1970) (*in Russian*).

103. E. Senogles, R.A. Thomas, *J. Polym. Sci., Polym. Symp.*, No.49, 203 (1975).

104. E. Senogles, R.A. Thomas, *J. Polym. Sci., Polym. Lett.Ed.*, **16**, No.11, 555 (1978).

105. E. Senogles, R.A. Thomas, *J. Polym. Sci., Polym. Symp.*, No.55, 241 (1976).

106. T.M. Karaputadze, V.I. Shumskii, Yu.E. Kirsh, *Vysokomol. Soed.*, **20A**, No.8, 1854 (1978) (*in Russian*).

107. B.V. Robinson, F.M. Sullivan, J.F. Borzella, S.L. Schwarz, *PVP. A Critical Rewiew of the Kinetics and Toxicology of Polyvinyl-Pirrolydone (Povidone). Michigan: Lewis Publ.,* 97 (1990).

108. Yu.E. Kirsh, *Poly-N,N-vinylpyrrolidine and others poly-N-vinylamides*, Moscow: Nauka (1998) (*in Russian*).

109. D.A. Topchiev, A.I. Martynenko, E.Yu. Kabanova, L.M. Timofeeva, *Vysokomol. Soed.*, **39A**, No.7, 1129 (1997) (*in Russian*).

110. D.G. Smith, *J.Appl. Chem.*, **17**, 339 (1967).

111. G.P. Gladyshev, *Polymerization of vinyl monomers*, Alma-Ata, 141 (1964) (*in Russian*).

112. V.A. Popov, G.P. Gladyshev, *Uspekhi khimii*, **42**, No.2, 273 (1973) (*in Russian*).

113. C.H. Bamford, E. Schofild, *Polymer*, **22**, 1227 (1981).

114. C.H. Bamford, E. Schofild, D.J. Michael, *Polymer*, **26**, 945 (1985).

115. I.Yu. Galaev, *Uspekhi khimii*, **64**, No.5, 505 (1995) (*in Russian*).

116. N.A. Plate, A.D. Litmanovich, O.V. Noa, *Macromolecule reactions*, Moscow: Khimiya (1977) (*in Russian*).

117. V.A. Kabanov, *Materials of the Third all-Russian scientific Kargin's conference "Polymers-2004"*, Moscow: MSU, **1**, 70 (2004) (*in Russian*).

118. R.V. Petrov, V.A. Kabanov, R.M. Khaitov, *Immunologiya*, No.1, 5 (1986) (*in Russian*).

119. T.V. Panova, E.V. Bykova, V.B. Rogachev, A.B. Zezin, *Materials of the Third all-Russian scientific Kargin's conference "Polymers-2004"*, Moscow: MSU, **1**, 340 (2004) (*in Russian*).

120. G. Moravetz, *Molecules in solution*, Moscow: Mir (1967) (*in Russian*).

121. R. Arnold, *J.Colloid. Sci.*, **12**, 549 (1957).

122. A. Silberberg, J. Eliassaf, A. Katschalsky, *J.Polymer Sci.*, **23**, 259 (1957).

123. H.P. Gregor, D.H. Gold, M. Frederick, *J.Polymer Sci.*, **23**, 467 (1957).

124. M. Mandel, J.C. Leyte, *J.Polymer Sci.*, **56**, 23 (1962).

125. J.C. Leyte, M. Mandel, *J.Polymer Sci.*, **A2**, 1879 (1964).

126. G. Barone, V. Grescenzi, F. Quadrifoglio, *Ricerca sci.*, **8**, 393 (1975).

127. T.N. Nekrasova, E.V. Anufrieva, A.M. El'yashevich, O.B. Ptitzyn, *Vysokomol. Soed.*, **7**, 913 (1965) (*in Russian*).

128. T.N. Nekrasova, E. Churylo, *Vysokomol. Soed.*, **11A**, 1103 (1969) (*in Russian*).

129. F.P. Grigor'eva, T.M. Birshtein, Yu.Ya. Gotlib, *Vysokomol. Soed.*, **9A**, 580 (1967) (*in Russian*).

130. F.P. Grigor'eva, L.A. Volkova, O.B. Ptitzyn, *Vysokomol. Soed.*, **11B**, 232 (1969) (*in Russian*).

131. F.P. Grigor'eva, L.A. Volkova, O.B. Ptitzyn, *Vysokomol. Soed.*, **12A**, 1363 (1970) (*in Russian*).

132. V.A. Kargin, S.Ya. Mirlina, V.A. Kabanov, G.A. Mikheleva, A.V. Vlasov, *Doklady AN SSSR*, **135**, 893 (1960) (*in Russian*).

133. L. Gargallo, D. Radic, *J. Polymer.*, **24**, 91 (1983).

134. W.G. Rothshild, *J. Am. Chem. Soc.*, **94**, No. 25, p.8676 (1972).

135. C. Tanford, V.G. Taggart, *J. Am. Chem. Soc.*, **82**, 6028 (1960).

136. D.R. Robinson, W.P. Jencks, *J. Biol. Chem.*, **238**, 1558 (1963).

Chapter 3. Radical copolymerization of guanidine containing monomers with diallyldimethylammonium chloride

3.1. Radical copolymerization with participation of acrylic acids

When considering kinetic effects found under investigation of radical polymerization of acrylate- and methacrylate-anions in water solutions in dependence on pH medium the conclusion was made about important contribution of chain propagation reaction to AA and MAA polymerization process due to formation of ion pairs at macroradicals ends that was conformed by direct measuring of individual kinetic constants [1]. To obtain information about the influence of mentioned electrostatic interactions on copolymerization reaction direction the AA and MAA copolymerization in water and organic solutions with a number of comonomers was studied.

Thus, Alfrey et. al. [2–4] realized copolymerization of N,N-diethylaminoethyl methacrylate (DEAEMA) with MAA and AA in water solutions. The results of determination of effective constants of DEAEMA and MAA copolymerization at various pH of reaction solutions are presented in Table 3.1.

Table 3.1. Effective constants of DEAEMA (**M₁**) with MAA (**M₂**) copolymerization

pH	r_1^{ef}	r_2^{ef}	$r_1^{ef} \times r_2^{ef}$
1,2	0,90	0,98	0,88
7,2	0,65	0,08	0,05

It is obvious, that with the rise of pH as far as methacrylic acid is ionized the sharp increase of units' tendency to alternating is observed. In this interval of pH in accordance with the values of ionization constants as DEAEMA (**pK_a** 10,0), so its units in copolymer are ionized. Authors assigned found effects to electrostatic factors, i.e. to the rise of probability of crossed propagation with pH increase at the expense of attraction of opposite charged particles.

Copolymerization of methacylate-anion and N-vinylpyrrolidone (N-VP) and also acrylamide (AAm) was carried out at 60°C in water solutions which pH was set by addition of NaOH (Table 3.2) [5, 6].

Table 3.2. Effective constants of MAA copolymerization with N-vinylpyrrolidone and acrylamide.

Comonomers	Medium	r_1^{ef}	r_2^{ef}
MAA – N-VP	water solution, pH 8	0,65 ± 0,02	0,33 ± 0,01
MAA – N-VP	2N water solution NaCl, pH 8	0,90 ± 0,04	0,30 ± 0,03
MAA – N-VP	water solution, pH 10	1,40 ± 0,02	0,30 ± 0,01
MAA – AAm	water solution, pH 8	0,25 ± 0,03	0,12 ± 0,02
MAA – AAm	2N water solution NaCl, pH 8	1,36 ± 0,02	0,18 ± 0,01

On the base of received data authors concluded that increase of effective constant $\mathbf{r_1^{ef}}$ with the rise of pH or at fixed pH = 8 with the rise of cation Na^+ content in solution mainly should be the result of increase of elementary constant $\mathbf{k_{11}}$, i.e. the rate constants of methacrylate-anion connection to "its" radical due to increase of probability if ion pair formation at the end of propagating chain. It is characteristic that the value of $\mathbf{r_2^{ef}}$ is practically not changed (as for N-VP, so for AA) that is the additional confirmation of constancy of reaction ability of mentioned comonomers and corresponding to them propagation radicals.

The same authors carried out investigations of radical copolymerization of AA and MAA with cation-active comonomer N,N-diallyl-N,N-dimethylammonium chloride (DADMAC) in water solutions [2–3, 5–7]. Kinetic measurements were made by dilatometer method. Sum monomers concentration in initial solution was 1,0 mole·l^{-1}, initiator concentration [APS] = 5·10^{-3} mole·l^{-1}, the temperature was 60°C. they established that DADMAC copolymerization with AA and MAA proceeded with noticeable rates in water solutions with formation of statistic copolymers of amorphous mature.

On the base of typical dependence of copolymer structure on the composition of initial monomer mixture under copolymerization of MAA with DADMAC in water solutions they revealed that at all initial molar ratios of comonomers the copolymer was enriched by units of acidic comonomer. With the help of these data the values of effective constants of copolymerization were determined by Faineman-Ross' and Mayo-Lewis' methods and they turned to be equal to: $\mathbf{r_1} = 0{,}01 \pm 0{,}0$; $\mathbf{r_2} = 2{,}0 \pm 0{,}4$ (where 1 – DADMAC; 2 – MAA) (pH was not fixed).

However, since AA and MAA homopolymerization rate is sharply reduced with the rise of pH under the transition from pH = 2 to pH = 5–6 (the rate of homopolymerization of DADMAC practically doesn't depend on pH), then under the change of mole part of acidic comonomer the pH of initial reaction solution (if it is not fixed during polymerization) is also changed significantly and due to this fact presented values of constants of copolymerization $\mathbf{r_1}$ and $\mathbf{r_2}$ are differed from constants obtained for the copolymerization system but with the fixed pH value.

In mentioned works [5–7] the results of investigation of radical copolymerization of DADMAC with AA and MAA in water solutions in the presence of APS at various (fixed values of medium pH are also presented. Kinetics of copolymerization of AA with DADMAC was investigated by dilatometry method at pH values of initial solution of comonomers 2 and 5 at various molar content of AA in initial monomer mixture in each case. In the case of pH = 2 with the rise of AA content in monomer mixture the polymerization rate was also increased. At pH = 5 this dependence is extreme and the rise of reaction rate is observed only in the interval $0 < [AA] < 0{,}7$ (in molar parts).

The calculation of copolymerization constants for monomer pairs (Table 3.3) DADMAC–AA and DADMAC–MAA allows concluding that in considered systems the most active comonomers are unsaturated acids.

Table 3.3. Effective constants of copolymerization in the system DADMAC ($\mathbf{M_1}$) : AA (MAA) ($\mathbf{M_2}$) (pH≈5)

Comonomers	$\mathbf{r_1}$	$\mathbf{r_2}$
DADMAC–AA	0,22 ± 0,13	3,44 ± 0,17
DADMAC–MAA	0,32 ± 0,13	5,2 ± 0,25

Authors also carried out kinetic measurements under copolymerization of DADMAC and AA (MAA) in acid, neutral and alkaline regions of pH in water solutions. Obtained data are presented in Table 3.4.

Table 3.4. The values of initial rates of copolymerization of DADMAC–MAA (AA) in acid, neutral and alkaline mediums (sum concentration of comonomers is 3 mole·l^{-1})

Copolymerization rate $\mathbf{V} \times 10^4$ (mole l^{-1} sec^{-1})							
DADMAC–AA				DADMAC–MAA			
The part of AA in mixture	pH 2	pH 6	pH 9	The part of MAA in mixture	pH 2	pH 6	pH 9
0,1	1,52	2,26	3,20	0,1	1,28	1,00	1,81
0,3	3,41	2,25	3,94	0,2	1,95	1,10	2,18
0,7	6,51	2,45	5,44	0,3	2,44	1,35	-

With the increase of acid molar part in initial monomer mixture obviously the copolymerization rate is naturally increased and for AA this rise is expressed for a greater extent than for MAA. The obtained data are in good agreed with obtained earlier results [6] on MAA and AA homopolymerization in water mediums. It was shown that under transition from acid mediums into neutral region of pH the initial rate of copolymerization was mainly decreased and the higher was the concentration of initial acid comonomer, the stronger was the mentioned decrease. On the contrary, under transition from neutral into alkaline region the naturally increase of the rate is observed that is in good agree with the data on the influence of pH on homopolymerization of AA (MAA) rate which were explained by V.A. Kabanov and D.A. Topchiev from position of ion pairs mechanism under polymerization of ionized monomers [6].

As it is obvious from previous sections, the most complete and systematic investigations of kinetic particularities and regularities of radical polymerization reactions of monomers of acrylic type in various solvents including water solutions were carried out in the 1960s–1970s.

Later on basing on the received earlier conceptions about the mechanism of radical polymerization reactions proceeding of unsaturated acids in mentioned systems the novel techniques of synthesis of polymers with another and sometimes unique properties were developed.

So, on the example of differing in chemical structure pairs of opposite charged PE in works by V.A. Kargin, V.A. Kabanov et.al. the equilibrium in reactions leading to formation of the novel wide class of polymer substances – inter-polyelectrolyte complexes (IPEC) was experimentally studied and quantitatively interpreted for the first time [8–12].

Creation of IPEC allows changing in desired way the charge, hydrophobicity and sizes of macromolecules, i.e. factors determining their basic physical-chemical properties that gives possibility of IPEC application in various branches of industry, agriculture, medicine. The principles and approaches to reception of novel polymers including those on the base of monomers of acrylic type using the method of matrix mechanism of macromolecules synthesis were developed [11–14]. These fields of high-molecular compounds chemistry are one of the most dynamically developing today [15–21] and the great number of publications devoted to this problem testifies to it.

At present the existence is proved and a number of properties is described for the novel class of inter-polymer complexes – three-component inter-polymer complexes with low-molecular intermediary formed by two polymer acids and low-molecular organic dibase [22–27]. Function of low-molecular intermediary is in realization of contacts between polymer components of polycation.

Among complex systems there are polyelectrolyte hydro-gels – hydrophilic polymer networks on the base of chemically cross-linked PE which possess unique physical-chemical properties: they may swell in water solutions and change their properties and volume in hundreds of times in dependence on solvent that allows considering their model objects as perspective materials for creation of medicines with prolonged action. Moreover in swollen state water interstice of hydrogel may be used for concentrating of low-molecular ions, organic dyes, medicines, PE macromolecules including DNA possessing the same charge as gel's chains [28, 29].

A lot of important particularities concerning behavior and properties of water-soluble non-stoichiometric polyelectrolyte complexes namely their ability to phase separation in narrow intervals of pH or poly-ions mixture structure change allow relating them to so-called "smart" or "intelligent" polymers. The interest to such polymers is grown from day to day due to perspectiveness of their use in biotechnology and medicine [28]. So, authors of [30–32] investigated properties of polyelectrolyte hydrogels with micro-porous structure on the base of sodium salts of PMAA and PDADMAC with the aim of their use for concentrating of low-molecular ions, organic dyes, medicines, polyelectrolytes macromolecules, including DNA. Moreover, they established that interaction of synthetic poly-cations and DNA in the form of complexes should be considered as perspective method of gene material delivery to the cell.

3.2. Radical copolymerization of diallyl guanidine monomers with diallyldimethylammonium chloride

As we noticed above (Section 2.1.2) DCT to monomer decisively influenced on DAGA and DAGTFA homopolymerization. At the same time it is worth saying that while studying DAGA with DADMAC radical copolymerization reaction the number of copolymers of various compositions with high intrinsic viscosities and consequently molecular masses was obtained [33–37].

Investigation of copolymerization processes of various comonomers except scientific interest of reveling of particularities of comonomers behavior under copolymerization proposes possibility of combination of positive in practical aspect features of separate monomers revealing in corresponding homopolymers. From this point of view study of radical copolymerization of synthesized by us DAGA with DADMAC – well known cationogenic monomer was of great interest.

Radical copolymerization of DADMAC with DAGA was carried out in vacuumed ampoules in water mediums ([**M**] = 4,0 – 4,3 mole·l^{-1}, initiator [APS] = 4×10^{-2} – 4×10^{-3} mole·l^{-1}, 60°C). Reaction solution of comonomers after addition of initiator was trice-degassed in vacuum up to 10^{-3} millimeter of mercury. Separated in acetone copolymer was filtered on glass filter, dried in vacuum and twice reprecipitated from methanol into ether. The copolymers composition was determined by ^{1}H NMR data (spectra were obtained in D_2O, the signal of methyl group of acetate counter-ion DAGA was the reference signal, see Section 4.2.3); intrinsic viscosity was measured by viscosimetry method in 1n water solution of NaCl at 30°C.

The wide set of copolymers of various composition in enough for further physical-chemical and biological investigations quantities was obtained as a result of carried out studies. Copolymerization was carried out up to various conversion degrees (2–93%) (investigation of polymerization and copolymerization up to deep conversion degrees may give important in practical meaning results).

In all cases including those at small conversions **q** ≤ 5% the formation of copolymers enriched by DADMAC units in comparison with initial mixture of comonomers is observed (Tables 3.5 and 3.6) that conformed high reaction ability of DADMAC in chain propagation reactions. We should us note that copolymerization reaction was observed only in the presence of radical initiators and was completely inhibited under introduction of effective radical inhibitors.

Table 3.5. The dependence of copolymer composition on initial structure of reaction solution under copolymerization of DADMAC (**M_1**) with DAGA (**M_2**)
([APS] = $4\cdot10^{-3}$ mole·l^{-1}, 60°C, copolymerization time was 40 hours)

No.	Initial mixture of comonomers		Copolymers		
	[**M**]$_{sum}$, mole·l^{-1}	**M_1:M_2**, mol. %	**M_1:M_2**, mol.%*	**q**, %**	[η],*** l·g^{-1}
1	4,00	20:80	30:70	2	0,14
2	4,00	40:60	56:44	3	0,19
3	4,00	50:50	68:32	6	0,21
4	4,26	70:30	93:7	18	0,39
5	4,17	80:20	97:3	31	0,48
6	4,12	90:10	98:2	52	0,64

Notes: * – was measured by ^{1}H NMR data; ** – conversion degree; *** – was determined at 30°C in 1n water solution of NaCl.

Table 3.6. The dependence of copolymer composition on initial structure of reaction solution under copolymerization of DADMAC (**M_1**) with DAGA (**M_2**)
([APS] = $4\cdot10^{-2}$ mole·l^{-1}, 60°C, copolymerization time was 33 hours)

No.	Initial mixture of comonomers		Copolymers		
	[**M**]$_{sum}$, mole·l^{-1}	**M_1:M_2**, mol. %	**M_1:M_2**, mol.%*	**q**, %**	[η],*** l·g^{-1}
1	4,00	20:80	38:62	10	0,10
2	4,00	40:60	75:25	23	0,13
3	4,00	50:50	83:17	28	0,17
4	4,29	70:30	87:13	74	0,20
5	4,17	80:20	91:9	80	0,31
6	4,17	90:10	94:6	93	0,52

Notes: * – was measured by ^{1}H NMR data; ** – conversion degree; *** – was determined at 30°C in 1n water solution of NaCl.

Thus, on the base of obtained data we may consider comparable reaction ability of considered comonomers that allows developing the methods of direct synthesis of novel cationogenic polymer products with prescribed characteristics (structure, molecular mass, hydrophilic-hydrophobic balance) that may be changed in a large diapason with consideration of this factor.

3.3. Radical copolymerization of guanidine monomer salts of acrylic acids with diallyldimethylammonium chloride in water mediums

Beginning from the seventies of the last century homo- and copolymers on the base of famous cationogenic monomer DADMAC are traditionally and widely used in many branches of industry also as flocculants, in nitrocellulose, ceramic and silicate production, under waste treatment, for precipitation of polymer latexes, stabilization of suspensions and food liquids. Moreover, they are used in electronics and production of contact lenses. In oil industry polymers on the base of DADMAC are used for destruction of oil emulsion, as oil antifreeze, during drilling process, for prevention of scale formation in oil-producing equipment and for some other aims. In agriculture they and also polymer complexes on their base serve for improvement of soils structure, for prevention of wind erosion. It is well known that (meth)acrylate (co)polymers possess wide range of practically useful properties; they may be used in pulp and paper, paintwork, textile industries as emulsifiers, for smoothing of synthetic fibers, for preparation of latexes, glue compositions and many others [38–44].

This list of applications is far from complete and represents only the small part of well known and widely used polymer mayterials on the base of monomers – DADMAC, AA and MAA.

Taking into account high biocide activity of guanidine containing compounds for a long time successfully applied in medicine and various fields of industry it was necessary to study the possibility of synthesis of novel copolymers on the base of guanidine containing monomers of acrylic type and DADMAC. Since it was reasonable to expect that newly created copolymers may reveal novel important properties and characteristics not inhering to initial homopolymers. (Part of results obtained by author of given book with collaborators was published in [45–49].)

Together with expecting practical meaning of mentioned polymers the study of kinetic particularities of radical copolymerization proceeding is undoubtedly relevant and useful in scientific respect first of all from the position of estimation of reaction ability of synthesized monomers at considered conditions.

The obtained in such way information is also necessary for effective control of processes of reception of copolymers with prescribed structures; distribution of chemical units in them and molecular-mass characteristics.

It is necessary to mention as it was established in this work that reception of PAG and PMAG by modification of initial polymer acids (PAA and PMAA) was impossible with 100% transformation in spite of seeming simplicity of given reaction.

So, under treatment of PAA and PMAA by guanidine even at conditions most favorable for modification (modification reaction was carried out in diluted 1% methanol solutions of polymers in homogeneous conditions, preliminary synthesized PAA and PMAA with low molecular masses ~ 40 000 – 60 000 at 3–5 repeated excess of guanidine and mixing for 5 days) the amount of connected with polymer chain guanidine groups was not higher than 50 mole %.

Under carrying of modification of corresponding copolymers of AA (MAA) with DADMAC by guanidine the final result will be more non-obvious due to possibility of proceeding of reaction between chlorine-anion of DADMAC and guanidine-cation with formation of GHC.

Generally speaking these results might be predicted a priory since as it was known reaction ability of functional groups of macromolecules was noticeably differed from that one for low-molecular analogous and the reason of this was chain nature of reagent with all consequent configuration, conformational and other effects [50] that was also revealed for given system. The presence of factors listed above serves as additional argument of necessity of carrying out of systematic kinetic investigations in copolymerization reactions of considering comonomers with the aim of creation on their base of polymer products with given properties and characteristics.

Kinetic measurements in copolymerization reactions were made with the use of dilatometry method. Copolymerization rate was determined by initial section of kinetic curve at conversion not higher than 5%. When calculating copolymerization rate we calculated the specific contraction for each composition basing on the values of specific contractions of corresponding homopolymers (lets remind that their values were 10,8% for GA and 7,0% fr MAG). The value of specific contraction for PDADMAC under polymerization in water solution was equal to 9,2%.

The optimal conditions of realization of given reactions – water mediums, sum concentration of copolymers [**M**] = 2 mole l^{-1}, [APS] = 5×10^{-3} mole l^{-1}, and 60°C were determined before carrying out of systematic kinetic investigations in considered by us copolymerization systems.

The studying of kinetics at given conditions showed that GA and MAG copolymerization reaction with DADMAC proceeded only in the presence of radical initiators and it was completely inhibited under introduction into reaction solution of effective radical inhibitor 2,2,6,6-tetramethyl-4-oxylpyridil-1-oxyl.

As a result of carried out kinetic investigations we established that rate of copolymerization in both systems DADMAC–AG and DADMAC–MAG was naturally reduced with the rise of content of diallyl monomer in initial mixture of comonomers (Figure 3.1), and for the first system (under polymerization with GA) the course o given dependence was expressed more clearly. The received results are in good agree with the data obtained in works [6, 7] under investigation of kinetics of copolymerization of DADMAC with AA and MAA in water solutions. In these systems they also established that copolymerization rate was decreased with the increase of DADMAC content in the initial reaction solution, and for AA this increase was expressed for a greater degree than for MAA (Table 3.7).

Table 3.7. The values of rates of copolymerization of DADMAC and (meth)acryl derivatives ([APS] = $5\cdot10^{-3}$ mole$\cdot l^{-1}$, H_2O, 60°C).

System, conditions	Copolymerization rate, $\mathbf{V_p}\times10^4$, mole l^{-1} sec^{-1}									
	Molar part of DADMAC in initial mixture									
	0,1	0,2	0,3	0,4	0,5	0,6	0,7	0,8	0,9	1,0
DADMAC–GA, $[\mathbf{M}]_{sum}$ = 2 mole$\cdot l^{-1}$; pH ≈ 7	10,5	7,4	-	5,5	5,2	5,0	-	3,8	2,7	1,2
DADMAC–MAG, $[\mathbf{M}]_{sum}$ = 2 mole$\cdot l^{-1}$; pH ≈ 7	2,3	2,1	2,3	2,0	2,6	2,2	1,9	2,2	2,0	1,2
DADMAC–AA, $[\mathbf{M}]_{sum}$ = 3 mole$\cdot l^{-1}$; pH ≈ 6	--	--	22,45	--	--	--	22,25	--	22,26	1,40
DADMAC–MAA, $[\mathbf{M}]_{sum}$ = 3 mole$\cdot l^{-1}$; pH ≈ 6	-	-	1,35	-	-	-	1,10	-	1,00	1,40

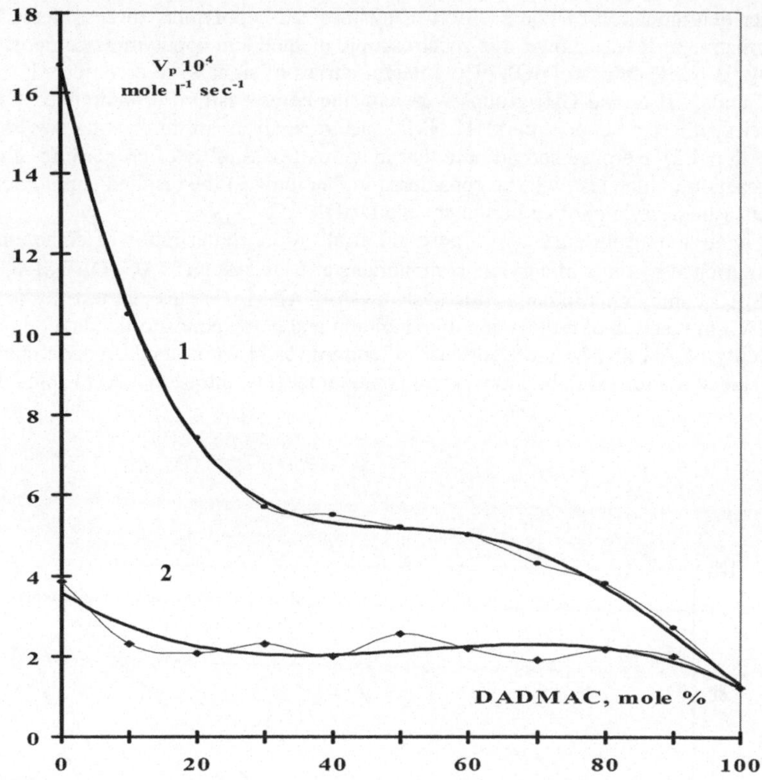

Figure 3.1. The dependence of initial rate of copolymerization on structure of comonomers mixture in the systems: DADMAC–GA (curve **1**) and DADMAC–MAG (curve **2**) $[\mathbf{M}]_{sum}$ = 2 mole l^{-1}; APS = $5 \cdot 10^{-3}$ mole l^{-1}; H_2O; 60°C.

At the same time the mentioned copolymerization systems – GA (MAG)–DADMAC and AA (MAA)–DADMAC are different.

So, under copolymerization of AA and MAA with DADMAC reaction solutions were homogeneous in the whole interval of compositions and formed copolymers were well dissolved in water and methanol. In our case complete homogeneity in copolymerization process was remained only in system DADMAC–GA that was apriory non-obvious since as it was established (Section 2.2.2.1) under GA homoplymerization homogeneity remained at concentration [GA] ≤ 1,3 mole l^{-1}.

In the system DADMAC–MAG homogeneity of reaction solution (at **q** ≤ 5%) was observed only at content [MAG] < 70% that corresponded to the [MAG] = 1,4 mole·l^{-1}. At higher content of MAG in reaction solution beginning from low conversions the turbidity was observed. Let us remained that under homopolymerization of MAG in water solutions the micro-heterogeneity of reaction solution was revealed at [MAG] > 0,4 mole^{-1}.

One more difference of considering copolymerization systems is in the following. Guanidine containing copolymers possessing good solubility in water were not diluted in organic solvents including methanol, DMSO, dimethylformamide.

Under determination of composition of synthesized copolymers in considered systems the most informative as it was shown was spectroscopic method and copolymers composition was determined by ^1H NMR data (in D_2O). (The integral curves of signals were compared; the signals of DADMAC units CH_2N and CH_3-groups were considered as basic). So, undoubtedly the data obtained under synthesized copolymers' ^1H NMR spectra analysis are interesting (we shall consider them in Section 4.2). Here we should note that in spite of possibility of proceeding of counter-ions exchange reaction (which also will be considered in Section 4.2) the resulted copolymers consist of acrylate units (methacrylate) of comonomers and DADMAC.

As it is obvious from curves in Figure 3.2 at all initial molar ratios of comonomers the copolymer is enriched by units of acrylate comonomers and the system MAG–DADMAC is more enriched by MAG comonomer in contrast to system GA–DADMAC. This fact testifies to higher reactivity of MAG in reaction of radical copolymerization and corresponds to the data on parameters of reaction ability of AA and MAA reported in literature [6, 7] where under copolymerization with DADMAC the MAA turned to be more active comonomer (in contrast to AA) (Table 2.8).

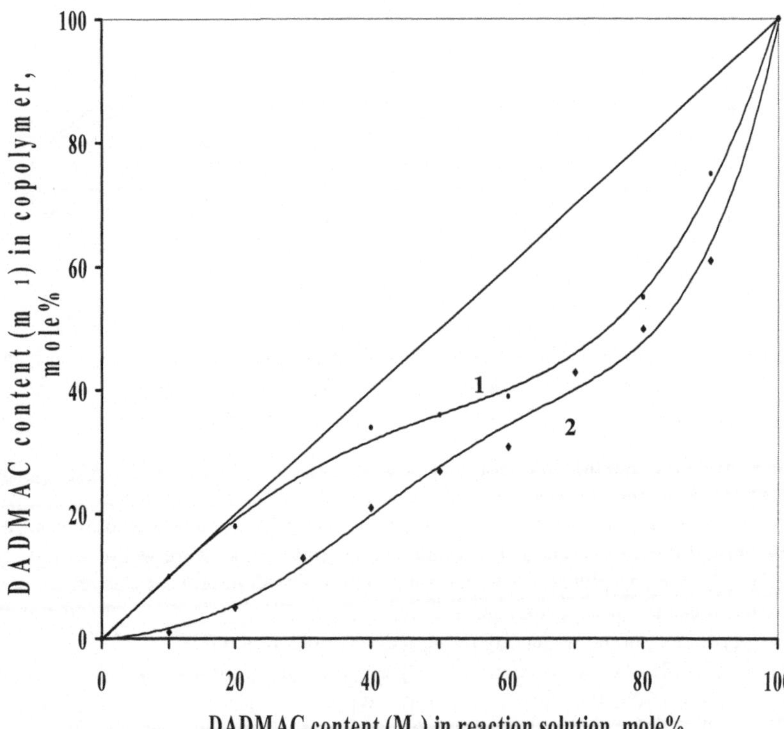

Figure 2.2. The dependence of structure of resulted copolymers in systems: DADMAC–GA (curve **1**) and DADMAC–MAG (curve **2**) on structure of initial reaction solution. $[\mathbf{M}]_{sum} = 2$ mole l^{-1}; APS = $5 \cdot 10^{-3}$ mole l^{-1}; H_2O; 60°C.

Table 3.8. The values of effective copolymerization constants in the systems DADMAC (M_1) –GA (M_2). $[M]_{sum}$ = 2 mole l^{-1}; APS = $5 \cdot 10^{-3}$ mole l^{-1}; H_2O; 60°C.

No.	Copolymerization systems	r_1	r_2	$r_1 \times r_2$
1	DADMAC–GA	0,20 ± 0,01	0,93 ± 0,02	0,186
2	DADMAC–MAG	0,10 ± 0,01	1,70 ± 0,03	0,170
3	DADMAC–AA, pH ≈ 5	0,22 ± 0,13	3,44 ± 0,17	0,757
4	DADMAC–MAKA, pH ≈ 5	0,32 ± 0,13	5,20 ± 0,25	1,664

For determination of copolymerization constants in binary system in practice various methods are used and the equation (3) of polymer structure underlies them [51]:

$$\frac{d[M_1]}{d[M_2]} = \frac{[M_1]}{[M_2]} \times \frac{r_1[M_1]+[M_2\}}{r_2[M_2]+[M_1\}} \qquad (3)$$

where $[M_1]$ and $[M_2]$ are the concentrations of monomers in initial mixture; r_1 and r_2 are copolymerization constants, $r_1=k_{11}/k_{12}$ and $r_2=k_{22}/k_{21}$.

Some methods may be used only in the case of low monomer conversions (not higher than 8%), in them the assumption is made that at the initial stage of copolymerization the values M_1 and M_2 are constant. That is why the ratio of rates of monomers consumption may be replaced by the ratio of molar concentrations of monomer units $[m_1]$ and $[m_2]$ in copolymer (equation 4):

$$\frac{[m_1]}{[m_2]} = \frac{[M_1]}{[M_2]} \times \frac{r_1[M_1]+[M_2\}}{r_2[M_2]+[M_1\}} \qquad (4)$$

They are for example the Mayo-Lewis' method of "lines crossing" [52], the analytical method of calculation of copolymerization constants [53], etc.

The methods of calculation of copolymerization constants which allow determination of structure of monomer mixture or copolymer practically at any monomer conversion are developed since equations of structure are calculated in integral form. The most simple of them is the Faineman-Ross' method [54].

Since we investigated copolymerization at conversion degrees, so for calculation of copolymerization constants we used analytical method, the values of constants calculated by this method were presented in Table 3.8 (experiments 1 and 2).

As it is obvious from these data in considered systems the preferred connection of (meth)acrylate guanidine to its radical is observed, that testifies to higher activity of given comonomers in reactions of radical copolymerization. The values of product $r_1 \times r_2 < 1$ indicates on pronounced tendency to alternation in both copolymerization systems.

High reactivity of MAG monomer in comparison with GA is probably caused by higher delocalization of carboxyl group charge in monomer molecule and shift of signals of MAG vinyl protons to the stronger field in comparison with GA in ^{1}H NMR spectra testifies to this fact (Section 4.2.2).

We should note that in both GA and MAG as we propose the energy of conjugation in monomer is higher than in the case of monomer acids AA and MAA.

It is undoubtedly interesting to carry out also comparable analysis of reactivity of synthesized by us guanidine containing salts of GA and MAG and initial AA and MAA monomers at identical conditions, i.e. to compare their rates of homopolymerization and copolymerization.

Unfortunately, it is impossible to carry out such quantitatively estimation in radical copolymerization reaction. Since in works on AA and MAA copolymerization with DADMAC the values of constants of relative activity were calculated by authors either at pH = 2, i.e. at conditions when acid monomer and corresponding to it units in comonomer were non-ionized, or in water medium without fixed value of pH (see Section 3.1).

The data on values of rates of homo- and copolymerization of corresponding monomers are presented in Table 3.7 and 3.9.

As it is obvious from kinetic data, the rates of GA and MAG homopolymerization are significantly higher than in polymerization at close conditions of corresponding monomer acids (the DAA was used as initiator for them which effectiveness was lower than for APS).

For synthesized in presented work copolymers the values of intrinsic viscosity [η] in a wide range of compositions were determined. As it is obvious (Figure 3.3) corresponding dependences have complex character and values of intrinsic viscosities of copolymers GA–DADMAC as well as the values of copolymerization rates are significantly higher than corresponding parameters of system MAG–DADMAC.

Table 3.9. The rate of homopolymerization of (meth)acrylate guanidine and (meth)acrylic acid in water solutions (pH ≈ 7)*

No.	Monomer	[**M**], mole l^{-1}	Neutralizing agent	Initiator, 5×10^{-3} mole l^{-1}	$V_p\times10^5$, mole l^{-1} sec^{-1}	Micro-heterogeneity
1	GA	1,0	-	APS	24,3	-
2	MAG	1,0	-	APS	14,0	+
3	AA [55]	1,2	NaOH	DAA	1,2	-
4		1,2	NH_4OH	DAA	2,0	-
5		1,2	TEA	DAA	1,3	+
6	MAA [55]	0,92	NaOH	DAA	0,95	-
7		0,92	NH_4OH	DAA	4,42	-
8		0,92	TEA	DAA	2,0	+

Notes: * – the rates for acids polymerization (experiments 3-8) were recalculated by values [AIBN] = 5×10^{-3} mole l^{-1}, since in Topchiev's work of reference polymerization was carried out at [AIBN] = 5×10^{-3} mole l^{-1}.

Thus, probably, the observing particularities and differences in a number of considered monomers are explained by complex character of contributions of various physical-chemical processes determining proceeding of homo-polymerization and copolymerization reactions of guanidine containing monomers of acryl type. At the same time the main contribution into the change of effective reactivity of polymerizing particles is made as we assume by associative interactions between guanidine and carboxyl groups (as intra-, so inter-molecular) and structural organization of corresponding monomers and polymers during copolymerization process.

Effect of solubility lost of macromolecules in reaction solution accompanying by appearance of micro-heterogeneity (which turned to be characteristic for considering polymerization system) is probably caused by the lack of water molecules. These molecules might be effectively used for macromolecules hydration but under introduction into water solution of polymer of "excessive" amount of guanidine groups the competition for water is appeared. This fact is observed under polymerization at high concentration of initial monomer (see Section 2.2.2.2). As a result of such "competition" in PAG (PMAG) macromolecule the displacement of water molecules from inter-chain space occurs and also from hydrated layers surrounding functional groups of monomer units (see also Figures 2.10 and 2.12). Chain is coagulated acquiring more compact conformation of "compressed" ball. Organic solvents (methanol, ethanol, dioxane) influence on macromolecule conformation in analogous way.

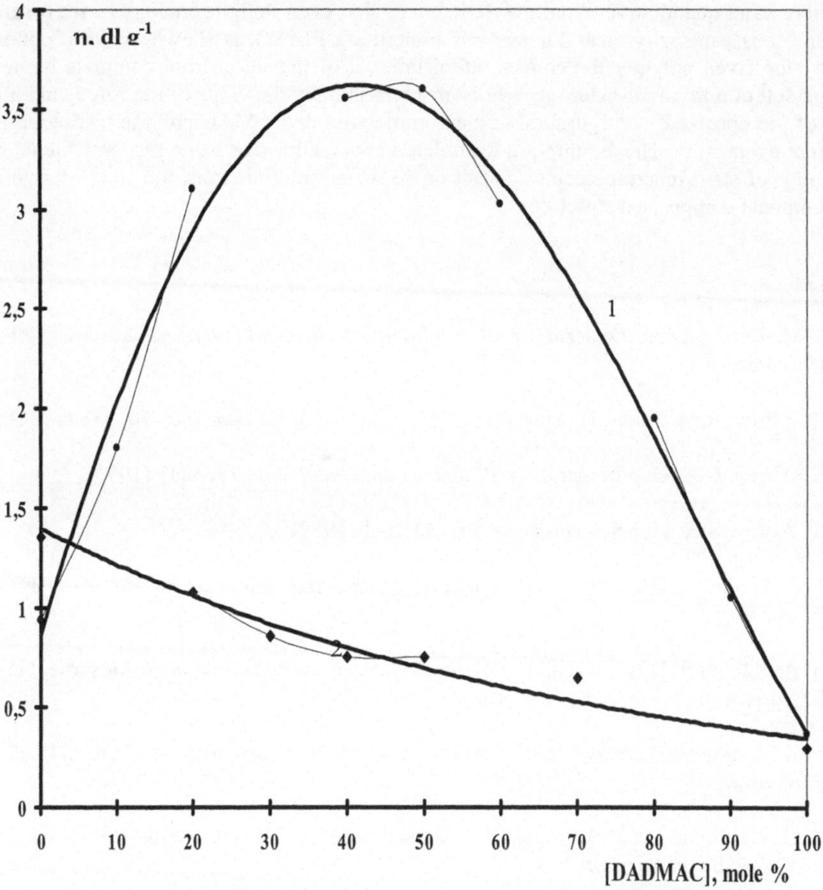

Figure 3.3. The dependence of intrinsic viscosity of copolymers on structure of initial reaction solution : **1** – DADMAC–GA, **2** – DADMAC–MAG. $[\mathbf{M}]_{sum.}$= 2 mole l^{-1}; 1n solution of NaCl in H_2O;30°C.

Such "salting-out" action of monomer "excess" as we showed was completely leveled (in the case of DADMAC–GA copolymerization) by introduction into copolymer of units of "neutral" comonomer (DADMAC). At the same time, maintenance of heterogeneity in system DADMAC–MAG at [MAG] ≥ 70 mol. % is probably caused by higher rigidity of copolymer of given structure that is caused by additional contribution of hydrophobic interactions of groups of MAG monomer.

A.R. Khokhlov et. al. created a lot of mathematical models of structures and spatial structures as for homopolyers, so for copolymers [56–58]. They formulated the idea about conformation–dependent design of primary consequence of copolymers containing hydrophilic and hydrophobic units and suggested the term "protein-like" copolymers for such structures [58] due to clear analogy in conformation transitions of analogous macromolecules by the type ball–globule. Of course, estimation methods of conformational transformations of considered by us polymerization systems carried out by viscosimetry method couldn't be the evidence of such transitions, one should use more strict quantitative criteria with this aim. However, definite analogy as we assume is obvious and it is especially clear for systems containing PMAG as shown above (Section 2.2.2.2). Namely for given polymer the concentration interval of transition from complete homogeneity of reaction solution to micro-heterogeneity is characteristic by the value of ion force, and also maintenance of "compressed" configuration of macromolecules under MAG polymerization until complete monomer exhausting. The assumption formulated above allows us using the term "denaturation" for describing of such macromolecules transition in water solution from the state of swollen ball to more compact compressed structure.

References

1. T.M. Karaputadze, *Dissertation of candidate of chemical sciences*, Moscow: INKhS (1972) (*in Russian*).

2. T. Alfrey, R.M. Fuoss, H. Morawetz, S.H. Pinner, *J. Am.Chem.Soc.*, **74**, 438 (1952).

3. T. Alfrey, C.G. Overberger, S.H. Pinner, *J. Am.Chem.Soc.*, **75**, 4221 (1953).

4. T. Alfrey, S.H. Pinner, *J. Polymer Sci.*, **23**, 553 (1957).

5. R.Z. Shakirov, *Dissertation of candidate of chemical sciences*, Moscow: INKhS (1973) (*in Russian*).

6. V.A. Kabanov, D.A. Topchiev, *Polymerization of ionized monomers*, Moscow: Nauka (1975) (*in Russian*).

7. A.I. Martynenko, *Dissertation of candidate of chemical sciences*, Moscow: INKhS (1981) (*in Russian*).

8. A.D. Narkevich, O.V. Kargina, V.A. Kabanov, V.A. Kargin, *Vysokomol. Soed.*, **12A**, No.8, 1817 (1970) (*in Russian*).

9. A.B. Zezin, V.V. Lutzenko, V.B. Rogachev, O.A. Oleksin, R.I. Kalyuzhnaya, V.A. Kabanov, V.A. Kargin, *Vysokomol. Soed.*, **14A**, No.4, 772 (1972) (*in Russian*).

10. A.D. Narkevich, O.V. Kargina, V.A. Kabanov, V.A. Kargin, *Vysokomol. Soed.*, **12A**, No.8, 1817 (1970) (*in Russian*).

11. I.M. Papisov, V.A. Kabanov, E. Osada, M. Leskano Brito, Zh. Reimont, A.N. Gvozdetzkii, *Vysokomol. Soed.*, **14A**, No.11, 2462 (1972) (*in Russian*).

12. I.M. Papisov, Tz.I. Nedyanova, N.K. Avramchuk, V.A. Kabanov, *Vysokomol. Soed.*, **15A**, No.9, 2003 (1973) (*in Russian*).

13. I.M. Papisov, A.A. Litmanovich, *Vysokomol. Soed.*, **19A**, No.4, 716 (1977) (*in Russian*).

14. V.A. Kabanov, I.M. Papisov, *Vysokomol. Soed.*, **21A**, No.2, 248 (1979) (*in Russian*).

15. A.B. Zezin, *Macromolecular reactions*, Moscow: Khimiya (1977) (*in Russian*).

16. V.A. Kabanov, *Materials of the Third Kargin's Conference "Polymers–2004"*, Moscow, **1**, 70 (2004) (*in Russian*).

17. R.V. Petrov, V.A. Kabanov, R.M. Khaitov, *Immunologiya*, No.1, 5 (1986) (*in Russian*).

18. V.A. Kabanov, *Macromol.Chem., Macromol. Symp.*, **1**, 101 (1986).

19. R.V. Petrov, V.A. Kabanov, R.M. Khaitov, A.V. Nekrasov, R.I. Ataullakhanov, *Allergy&Clinical Immunology*, **15**, 56 (2003).

20. I.M. Okhapkin, E.E. Makhaeva, A.R. Khokhlov, *Materials of the Third Kargin's Conference "Polymers–2004"*, Moscow, **1**, 120 (2004) (*in Russian*).

21. *Materials of the First international Symposium on Polyelectrolytes.* Potsdam, Germany, 17, 73-75, 78, 79, 151 (1995).

22. O.V. Kargina, O.V. Prazdnichnaya, N.V. Avramenko, M.N.Frolova, S.L. Davydova, *Vysokomol. Soed.*, **35A**, No.10, 1611 (1993) (*in Russian*).

23. O.V. Kargina, O.V. Prazdnichnaya, S.V. Korableva, I.N. Kuz'min, *Vysokomol. Soed.*, **36A**, No.8, 1316 (1994) (*in Russian*).

24. O.V. Kargina, O.V. Prazdnichnaya, I.D. Yurgens, E.Yu. Badina, *Vysokomol. Soed.*, **39A**, No.1, 22 (1997) (*in Russian*).

25. O.V. Kargina, O.P. Komarova, G.N. Bondarenko, *Vysokomol. Soed.*, **44B**, No.12, 2232 (2002) (*in Russian*).

26. N.I. Aksenova, A.V. Kharenko, A.B. Zezin, G.B. Bravova, V.A. Kabanov, *Vysokomol. Soed.*, **40A**, No.3, 403 (1998) (*in Russian*).

27. V.A. Izumrudov, San Khyun Lim, *Vysokomol. Soed.*, **40A**, No.3, 459 (1998) (*in Russian*).

28. I.Yu. Galaev, *Uspekhi khimii*, **64**, No.5, 505 (1995) (*in Russian*).

29. *Materials of the First international Symposium on Polyelectrolytes.* Potsdam, Germany, 30, 85, 159, 160 (1995).

30. *Materials of all-Russian Kargin's symposium "Chemistry and physics of polymers in the beginning of XXI century"*, Chernogolovka, 3-16, 3-52, 4-37 (2000) (in Russian).

31. O.V. Nikolaeva, T.V. Budtova, L.M. Kalyuzhnaya, N.G. Bel'nikevich, E.N. Vlasova, S.Ya. Frenkel', *Vysokomol. Soed.*, **41A**, No.7, 1176 (1999) (*in Russian*).

32. O.V. Nikolaeva, T. Budtova, Yu. Brestkin, Z. Zoolshoev, S. Frenkel, *J. Appl. Polym. Sci.*, **72**, 1523 (1999).

33. G.E. Zaikov, Yu.A. Malkanduev, S.Yu. Khashirova, A.M. Esmurziev, A.I. Martynenko, L.I. Sivova, N.A. Sivov, *in Biochemistry and Chemistry: Research and Developments*, Nova Science Publishers, New York, 39 (2003).

34. G.E. Zaikov, Yu.A. Malkanduev, S.Yu. Khashirova, A.M. Esmurziev, A.I. Martynenko, L.I. Sivova, N.A. Sivov, *in Chemical Reactions in Liquid and Solid Phase: Kinetics and Thermodynamics*, Nova Science Publishers, New York, 165 (2003).

35. G.E. Zaikov, Yu.A. Malkanduev, S.Yu. Khashirova, A.M. Esmurziev, A.I. Martynenko, L.I. Sivova, N.A. Sivov, *Russian Polymer News*, **8**, No. 4, 1 (2003).

36. G.E. Zaikov, Yu.A. Malkanduev, S.Yu. Khashirova, A.M. Esmurziev, A.I. Martynenko, L.I. Sivova, N.A. Sivov, *J. Environ. Protect. And Eco*logy, **4**, No. 4, 863 (2003).

37. G.E. Zaikov, Yu.A. Malkanduev, S.Yu. Khashirova, A.M. Esmurziev, A.I. Martynenko, L.I. Sivova, N.A. Sivov, *J.Appl.Pol.Sci.*, **91**, 439 (2004).

38. N.M. Boyarkina, V.V. Kryuchkov, E.S. Parkhamovich, L.A. Amburg, D.A. Topchiev, V.A. Kabanov, *Plast. Massy*, No. 8, 17 (1987) (*in Russian*).

39. Inventors certificate 1578083 USSR B.I. 1990, No.26.

40. Inventors certificate 1252329 USSR B.I. 1986, No.31.

41. Inventors certificate 1129215 USSR B.I. 1984, No.46.

42. Inventors certificate 1401813 USSR B.I. 1988, No.21.

43. Inventors certificate 865887 USSR B.I. 1981, No.35.

44. G.B. Butler, *Cyclopolymerization and Cyclocopolymerization*, Marcel Dekker, Inc., New York, 558 (1992).

45. N.A. Sivov, A.I. Martynenko, E.Yu. Kabanov, N.I. Popova, G.N. Bondarenko, S.Yu. Khashirova, A.M. Esmurziev, *Materials of scientific conference INKhS RAS*, Moscow, 63, 69 (2003) (*in Russian*).

46. N.A. Sivov, A.I. Martynenko, E.Yu. Kabanov, N.I. Popova, A.M. Esmurziev, *Materials of the Third Kargin's Conference "Polymers–2004"*, Moscow, **1**, 168 (2004) (*in Russian*).

47. N.A. Sivov, A.I. Martynenko, E.Yu. Kabanov, N.I. Popova, S.Yu. Khashirova, A.M. Esmurziev, *Neftekhimiya*, No.1, 47 (2004) (*in Russian*).

48. A.M. Esmurziev, M.P. Filatova, A.N. Sivov, G.N. Bondarenko, M.A. Dzyubina, A.I. Martynenko, N.A. Sivov, *Materials of all-Russian conference "Structureand dynamics of molecular systems"*, Yal'chik, 343 (2003); 312 (2004) (*in Russian*).

49. N.A. Sivov, A.I. Martynenko, E.Yu. Kabanova, L.I. Sivova, G.E. Zaikov, *J.Appl.Pol.Sci.*, 2005, *in press*.

50. N.A. Plate, A.D. Litmanovich, O.V. Noa, *Macromolecular reactions*, Moscow: Khimiya (1977) (*in Russian*).

51. *Copolymerization*, Ed. by D. Khema, Moscow: Khimiya (1971) (*in Russian*).

52. T. Alfrey, G. Goldfinger, *J. Chem.Phys.*, **12**, N 6. 205 (1944).

53. A.I. Ezrielev, E.L. Brokhina, B.S. Roskin, *Vysokomol. Soed.*, **11A**, No.8, 1670 (1969) (*in Russian*).

54. M. Fineman, S.D. Ross, *J.Pol.Sci.*, **5**, 251 (1950).

55. D.A. Topchiev, *Dissertation of doctor of chemical sciences*, Moscow: INKhS (1973) (*in Russian*).

56. A.R. Khokhlov, P.G. Khalatur, *Physica A.*, **249**, 253 (1998).

57. A.R. Khokhlov, P.G. Khalatur, *Phys. Rev. Lett.*, **82**, 3456 (1999).

58. V.I. Lozinskii, I.A. Symenel, E.A. Kurskaya, V.K. Kulakova, V.Ya. Grinberg, A.S. Dubovik, I.Yu. Galaev, B. Mattiasson, A.R. Khokhlov, *Doklady RAS*, **375**, No.5, 637 (2000) (*in Russian*).

Chapter 4. Investigation of physical and chemical properties of guanidine-containing monomers and (co)polymers

4.1. Synthesized compounds and their IR spectroscopy characteristics

In given Section the IR-spectral characteristics and their correlation with structure of synthesized derivatives of diallyl guanidine: DAGA, DAGTFA, guanidine monomer salts GA and MAG and also homopolymers PAG, PMAG and copolymers of DAGA and DADMAC, guanidine acrylate salts and taken as model compounds guanidinehydrochloride (GHC), triguanidine phosphate (GP) and also PDADMAC are considered. IR spectra of all compounds were registered in solid type in tablets of KBr.

The analysis of IR-spectral characteristics was made by comparing of monomer guanidine containing salts with their structure analogous taken as models and also by comparing of spectra of monomer and polymer compounds which should confirm corresponding changes in spectra under transition from monomers to (co)polymers. For construction of counter-ion nucleus (in both monomers and polymers) we may propose two stable structures with various type of bonding (on the example of GA and MAG) (Scheme 4.1): at the expense of hydrogen bonding of charged amino-group with delocalized acrylate-ion (the first type of bonding), or with participation of delocolized systems of two nitrogen atoms and acrylate counter-ion (the second type of bonding).

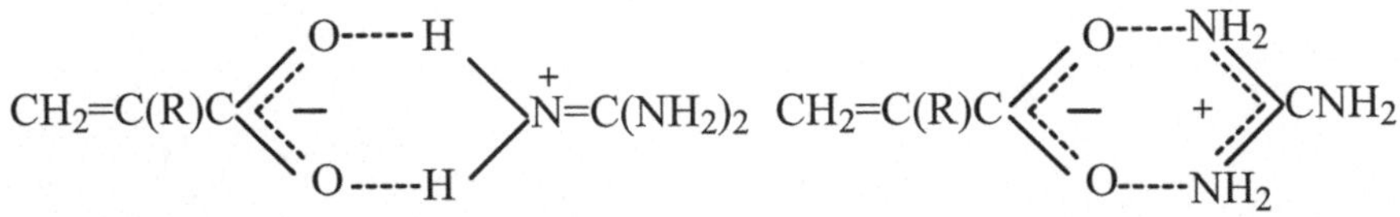

the first type of bonding　　　　the second type of bonding

Scheme 4.1.

IR-spectral investigations confirmed the structure of monomer and polymer products obtained in presented work.

So, in MAG monomer spectrum (Table 4.1, Figure 4.1) there are two broad but clearly expressed bands at 3385 and 3100cm^{-1} which relate to stretching vibrations of N-H bonds in guanidine cation.

Two intensive bands at 1680 and 1656cm^{-1} as in the case of other salts of guanidines (GHC and GP) are connected with N=C stretching vibrations and mixed with them deformations of NH_2 groups. Two bands at 520 and 544cm^{-1} relating to angle deformations of CNH are also observed. In other words all bands characteristic for guanidine structure present in MAG spectrum, very close bands were observed in spectra of model GHC and GP. Methacrylate counter-ion also revealed in spectrum in the shape of typical bands characterizing carboxylate-ion of methacrylic acid. Broad and very intensive band at 1528cm^{-1} relates to stretching vibrations of bond C=O in the anion –COO^{-}. In the spectrum of sodium salt of methacrylic acid the analogous band lies at 1540–1560cm^{-1} and since the position and intensity of this band are determined by delocalization degree of electrons in group 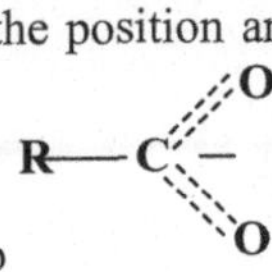 we may assume that in the case of MAG delocalization of electrons by bonds C–O in methacrylate-anion should be very high.

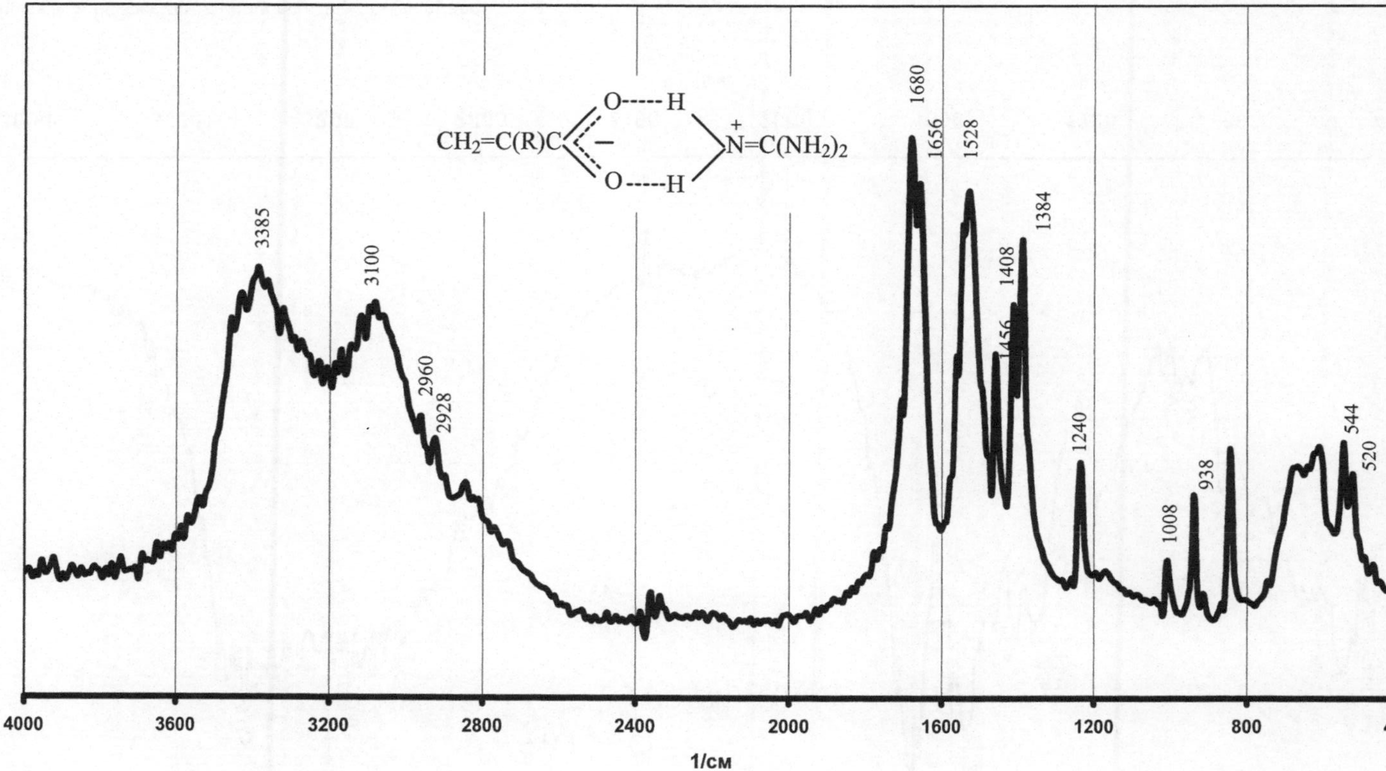

Figure 4.1. IR-spectrum of methacrylate guanidine.

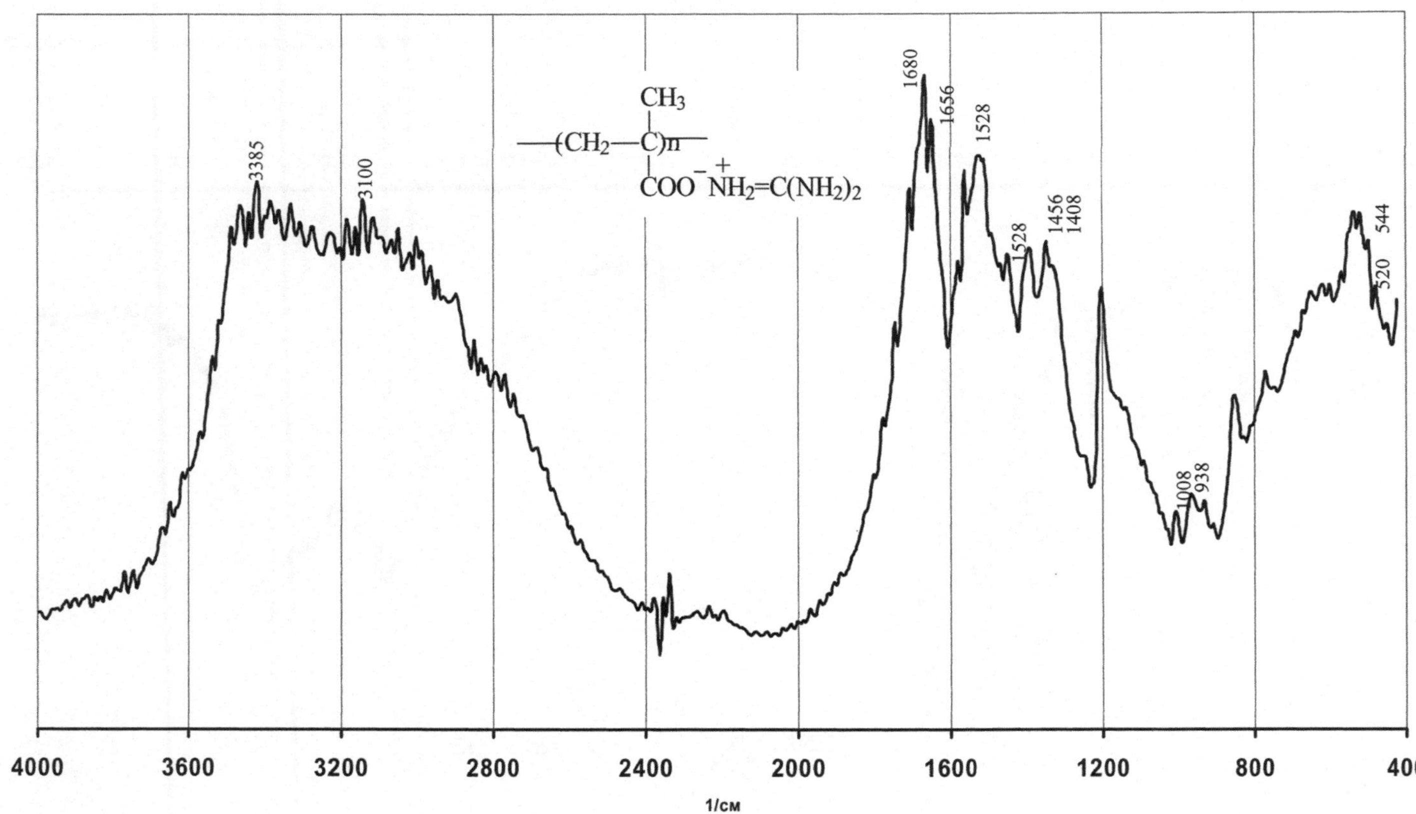

Figure 4.2. IR-spectrum of polymethacrylate guanidine.

Table 4.1. IR-spectral data of acrylic derivatives of guanidine*

$CH_2=C(R)COO^- \, ^+H_2N=C(NH_2)_2$

AG: R=H; MAG: R=CH_3

Monomer	Guanidine fragment			
	ν (NH) stretching	ν (C=N) stretching	ν (NH_2) deformational	ν (CNH) angular deformation
MAG	3100, 3385	1680	1656	520, 544
GA	3091, 3418	1674	1660	529, 544
Monomer	Vinyl fragment			
	ν (CH) stretching	ν (C=O) stretching	ν (RC=) skeletal deformation	ν (CH_2=C-) non-planar deformation
MAG	2928, 2960	1528	1240, 1384, 1408, 1456	938, 1008
GA	2929, 2960	1524	1275, 1359, 1419	956, 988

* – Position of peaks of corresponding signals are presented in cm^{-1}

The other bands characterizing methacrylate anion are also well revealed in spectrum: 2960, 2928 cm^{-1} (stretching vibrations of CH bonds), 1456, 1408, 1384, 1240 cm^{-1} – skeletal deformational vibrations in the group CH_3-C=. The most important are the bands at 938, 1008cm^{-1} characteristic for non-planar deformational vibrations of CH_2=C- in methacrylates. Since the bands of stretching vibrations of double bond in the region 1640–1660cm^{-1} and stretching vibrations ν CH at double bond (3000–3080 cm^{-1}) are closed by strong absorption in these regions of guanidine fragment, then the bands of deformational vibrations of CH_2 at double bond mentioned above remain the only spectral test for observing the disappearance of double bonds of methacrylate under polymerization. Indeed, if one compares spectra of monomer MAG and polymer PMAG (Figure 4.2) he will see that qualitatively these spectra are slightly differed, namely the bands at 938 and 1008cm^{-1} are sharply decreased in their intensity that testifies to the fact that double bonds in this compound are practically absent. The presence of these bands although of low intensity indicates that polymer has not high molecular mass and in spectra we can see and groups (actually, [η] of given sample of homopolymer was 0,2dl·g^{-1}). Qualitative coincidence of bands of both guanidine fragment (NH_2 – 3450 – 3100 cm^{-1}, C=N – 1660 – 1680 cm^{-1}), and carboxylate-ion (1560 – 1520 cm^{-1}) in monomer and polymer spectra is the guarantee of the fact that nature of ion bonding in polymer and monomer is similar. Analogous regularities were observed for GA (Table 4.1) and PAG.

On the base of these data we may conclude that in acrylate derivatives as in monomer, so in polymer ones the first type of counter-ion bonding is realized. The suggestion about possibility of realization of various types of bonding was proposed in accordance with the data obtained for DAGA and DAGTFA. Comparison of spectra of DAGA (Figure 4.3) and DAGTFA (Figure 4.4) demonstrates significant differences in spectra of these would seem isostructural compounds. The fact that in trifluoroacetate spectrum there are intensive bands 1200 – 700cm^{-1} which may be connected with stretching and deformational vibrations in CF_3 group and in acetate spectrum such bands are absent is not the main difference. In trofluoroacetate spectrum the band of guanidine fragment are revealed in the same way, as in spectra of other guanidine salts described above. In other words, N–H stretching vibrations have two intensive broad bands with maximums at 3335 and 3160cm^{-1}, and stretching N=C vibrations together with deformation of HNH angle have broad and very intensive band with maximum at 1666cm^{-1}. Vibration of C=O band in carboxylate counter-ion is revealed at 1598cm^{-1} that indicates on low delocalization degree of electrons in such counter-ion in comparison with methacrylate counter-ion. There is quite another picture in this region of spectrum of acetate. The very intensive band at 1579cm^{-1} is shifted by 19cm^{-1} to the region of long waves in comparison with trofluoroacetate spectrum, i.e. delocalization degree in group O-C=O in this case is higher, than in trifluoroacetate counter-ion. But the main thing is that there are no intensive bands of N=C bonds in the region 1660 – 1680 cm^{-1} in DAGA spectrum. The bands of stretching vibrations of N–H bonds are strongly shifted to the region of long waves (3210 and 3009 cm^{-1}) and are weaker by intensity. So, basing on spectral features we may assume that natures of bonding of counter-ions in DAGA and DAGTFA are various; for the first one we should expect the second type of bonding (structure IV, scheme 2.5, section 2.1.2), and for the first one as well as for all other guanidine derivatives described above – the second type of bonding (structure V, scheme 2.5, section 2.1.2).

In IR-spectra of compounds having allyl substituent 4 characteristic bands ν =CH (3080 – 3040 cm^{-1}), ν C=C (1640 – 1660 cm^{-1}) and two bands of deformational vibrations =CH_2 –990 – 1010 cm^{-1} and 910 – 930 cm^{-1} are usually observed, and the last one usually has very high intensity. The presence of nitrogen atom in allyl position to double bond usually doesn't reduce the intensity of given bands, in any case in triallyl amine, methyldiallyl amine and dimethylallyl amine spectra the band in the region of 920cm^{-1} remains the most intensive band in spectra. In spectra of guanidine allyl derivatives the bands of stretching vibrations of bonds CH and C=C are usually closed by intensive bands of guanidine groups, but bands in the region 910 – 1000 cm^{-1} should be clearly observed. Indeed, in DAGA spectrum there are bands at 925, 1011 cm^{-1} and even the band at 1647 cм$^{-1}$ which may be assigned to stretching vibrations ν C=C, however the intensity of band 925 cm^{-1} is weaker in comparison with intensities of analogous bands in allylamines.

In DAGTFA spectrum the band at 910cm^{-1} is revealed as arm of band at 930cm^{-1} and its intensity is too low, the rest bands characteristic for allyl substituent are not revealed at all. In other words interaction of protons at double bond with some fragment of compound occurs, however it is not simply the delocalization with nitrogen electron pair.

IR-spectrum of statistic copolymer DADMAC with DAGA (Figure 4.5) has a number of interesting particularities. The spectrum of DAGA monomer in which there are no bands of C=N bonds is described above. In copolymer spectrum there is intensive band with maximum at 1648 cm^{-1} which of course is distorted by absorption of water deformational vibrations in this region, but their intensity and several inflections on arms testify to the fact that in given compound there are N=C and NH_2 and H_2O bonds. The absorption band of acetate counter-ion is slightly shifted to the region of long waves in comparison with analogous band in monomer spectrum: 1560cm^{-1} (1579cm^{-1} in monomer). In other words, we may assume that in DAGA monomer the second type of counter-ions bonding is realized, and in copolymer – the first one. The bands of DADMAC units are weakly revealed in copolymer spectrum, and bands of vinyl groups which are usually observed in PDADMAC are absent in copolymer spectrum that is probably indicates on the absence of effective chain transfer on comonomer DADMAC as a result of which the vinyl groups are appear.

Finally, IR-spectra of statistic copolymers DADMAC with GA and MAG show that all polymer compounds maintain ion character (there is carboxylate-anion band in all spectra), but there is one particularity. The bands of DADMAC units are weakly but revealed in copolymers spectra, and bands of vinyl groups which are usually obvious in PDADMAC spectrum are revealed in copolymers spectra due to the presence in copolymers structure of counter-ions of diallyldimethylammonium at acrylate units. As we mentioned above, under DADMAC copolymerization for example with DAGA there was no effective chain transfer on comonomer DADMAC as a result of which vinyl groups appeared.

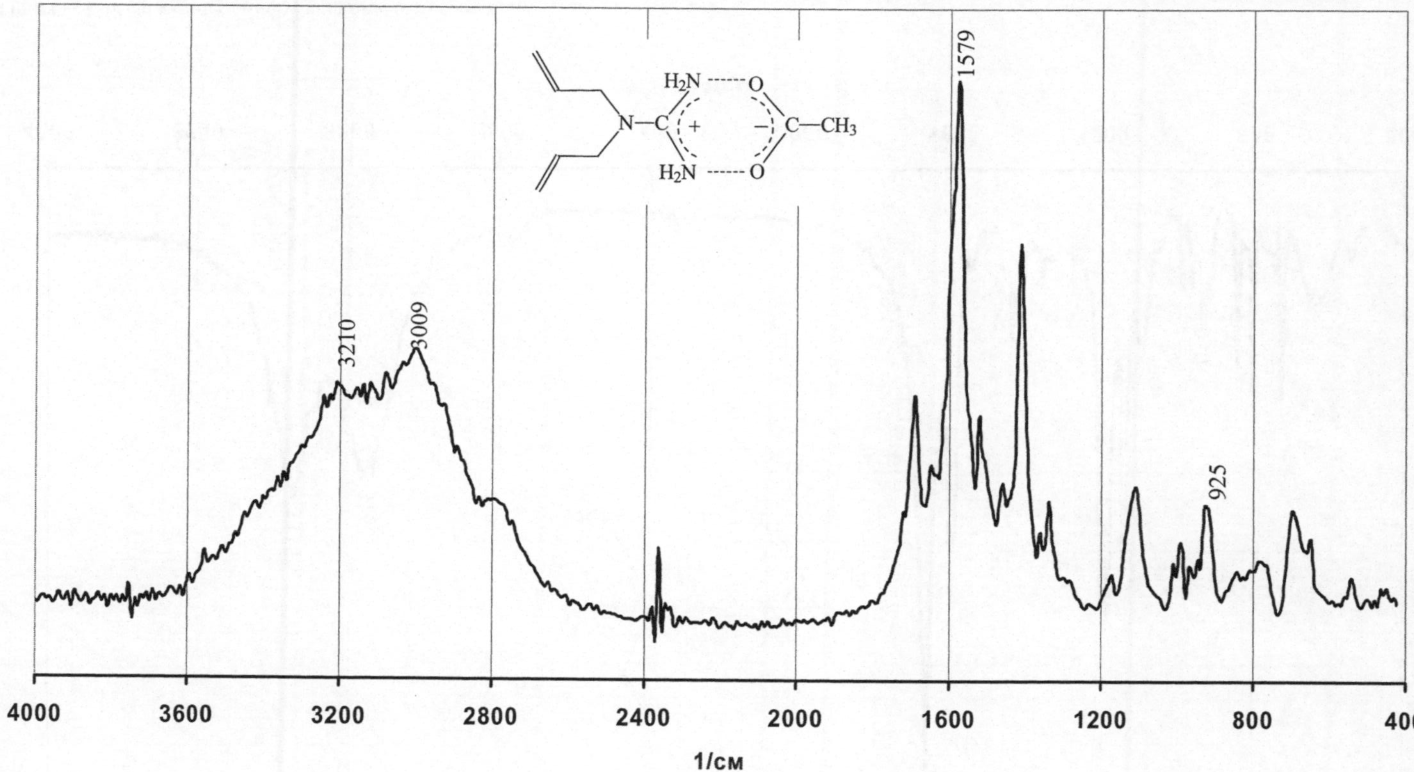

Figure 4.3. IR-spectrum of diallylguanidine acetate.

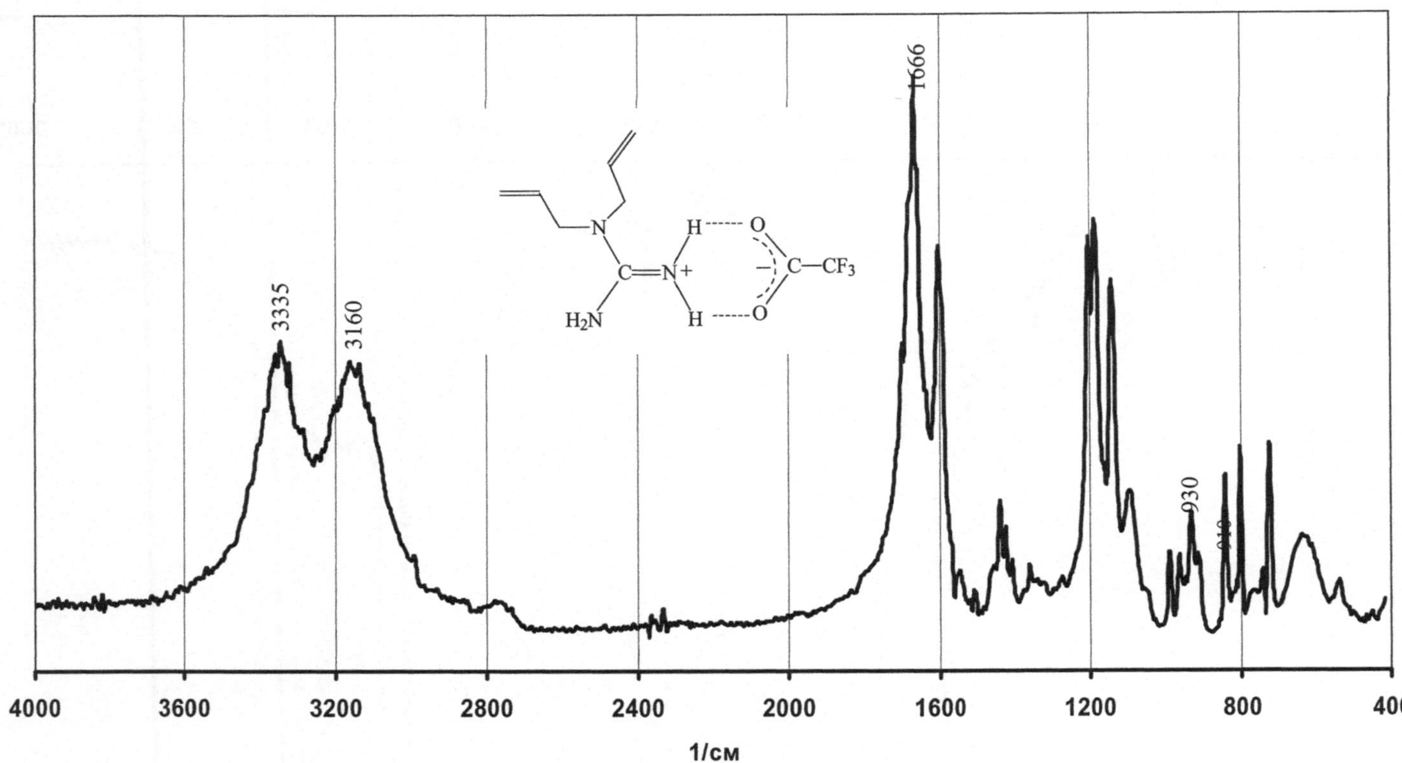

Figure 4.4. IR-spectrum of diallylguanidine trifluoroacetate.

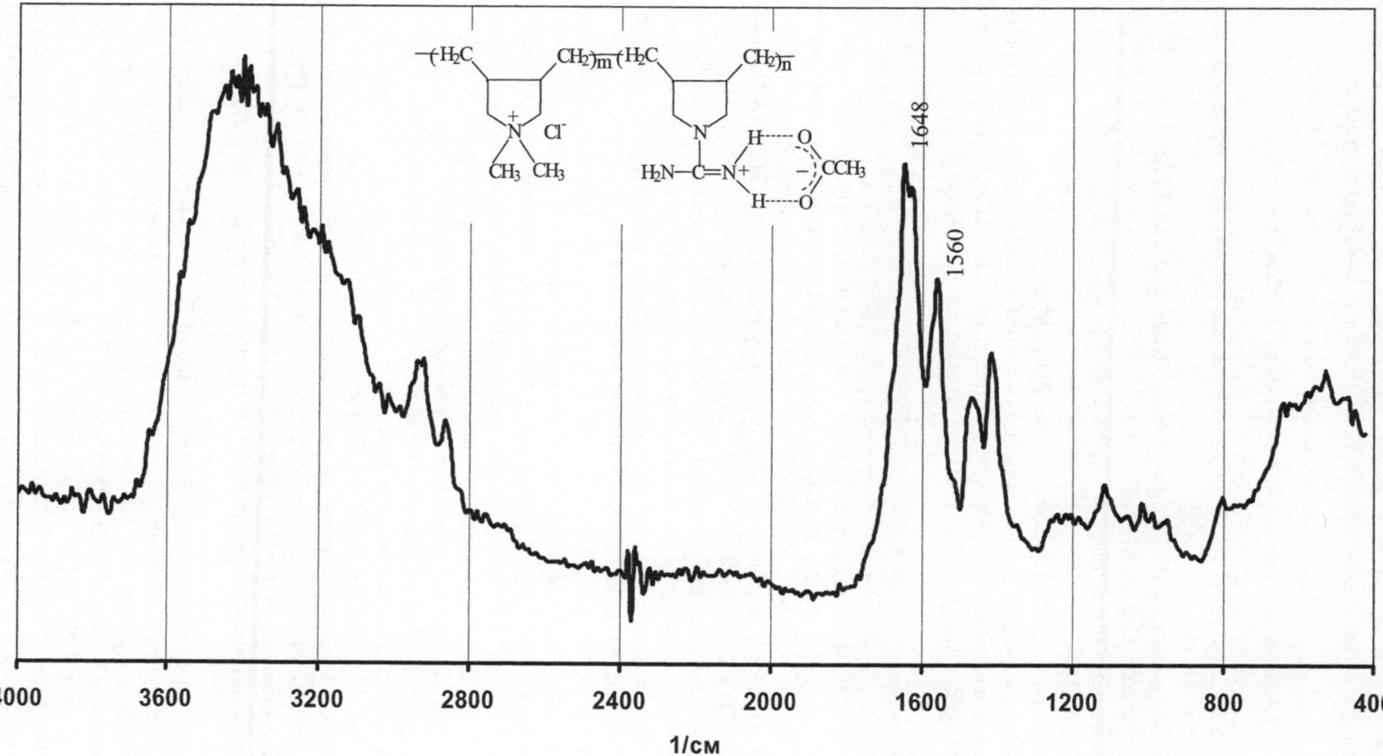

Figure 4.5. IR-spectrum of diallyldimethylammonium chloride copolymer with diallylguanidine acetate (75 : 25)..

As a conclusion we should note, that ^1H NMR spectroscopy which data will be considered in the following Section turned to be more informative in determination of copolymers structure.

4.2. Investigation of synthesized monomer and polymer products by ^1H NMR spectroscopy

In presented Section we analyze ^1H NMR spectral characteristics and their correlation with structures of the following compounds:

- Synthesized guanidine containing diallyl derivatives of DAGA and DAGTFA;
- Derivatives of acrylic acids GA and MAG;
- Their homopolymers PDAGA and PDAGTFA, PAG and PMAG;
- Copolymers with DADMAC;
- One of DADMAC copolymers and its homopolymer PDADMAC;
- A number of model compounds close in structure to studied in work monomers and polymers and assisting in understanding of basic regularities of structure and behavior in various mediums of novel synthesized guanidine containing monomers and polymers.

4.2.1. Diallyl derivatives

In given Section ^1H NMR spectral characteristics and their correlation with the structures of synthesized derivatives of diallylguanidine: namely of DAG, DAGA and DAGTFA; and also of DADMAC taken as model compounds of diallyl amine (DAA – the initial substance under DAG production) and synthesized earlier in laboratory of polyelectrolytes chemistry IPCS RAS diallylmethylamine (DAMA) and its trifluoroacetate (DAMATFA) are considered.

The analysis of signals character, their chemical shifts and integral intensities completely conforms assumed structure for given compounds. The main spectral characteristics of signals of various groups of protons of investigated by us substances are summarized in Table 4.2; ^1H NMR spectra of DAGA, DAGTFA, DADMAC monomers are presented in Figures 4.6–4.9.

The differences in spectra of these substances allow concluding the following about their structure and behavior in solvents in which spectra were measured. Spectra of allyl parts of these compounds relate to the type of $ABMX_2$–systems leveled to a various degree (signals of methyl groups of DAGA, DADMAC and DAMA are revealed as singlets). Comparison of positions of protons signals of group CH_2-N of allyl system of DAG and DAA (Table 4.2, experiments 3, 7) indicates on dis-screening action of guanidine group (we should note that this conclusion will be very important when analyzing spectral data of guanidine containing acrylate derivatives, Section 4.2.2). Protonation of guanidine substituent leads to higher shift of signal of CH_2-N group to the weak field – 4,09 ppm for DAGA and 4,31 ppm for DAGTFA (Table 4.2, experiments 1 and 5). At that we should consider also strong –I–effect of CF_3 group in contrast to +I–effect of methyl group that leads to differences in chemical shifts of protons of given group for DAGA and DAGTFA. Probably the differences in forming structures IV for DAGA and V for DAGTFA are explained by the same fact (scheme 2.5, Section 2.1.2) that was confirmed by IR-spectroscopy method for various monomer salts.

Comparison of DAGA spectra measured in DMSO-d_6 and of DAG measured in water (Table 4.2, experiments 2 and 4) indicates on closeness of chemical shofts of protons of all groups of allyl system for these compounds. It may be explained by the fact that in less polar DMSO for structure IV delocalization of charges in system is higher than in water that leads to sharp decrease of dis-screening influence of positively charged guanidine group and shift of signals to stronger field. Analogous picture may be observed for DAGTFA in deutero-acetone (Table 2.4, experiment 6); it is less clearly expressed since there is influence of CF_3 group at the expense of strong –I–effect. At that we observed leveled multiplicity of signals of CH_2-N and CH_2= groups (Table 4.2, experiments 1–6).

Table 4.2. Spectral characteristics of diallyl derivatives*, **

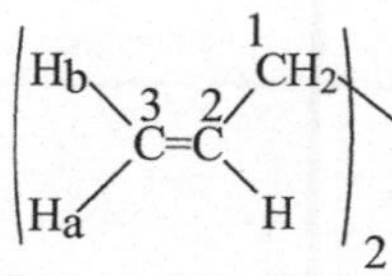

$X = NC(NH_2){=}\overset{+}{N}H_2\ CH_3COO^-$ ***DAGA*** $X = NC(NH_2){=}\overset{+}{N}H_2\ CF_3COO^-$ ***DAGTFA***

$X = NC(NH_2){=}NH$ ***DAG*** $X = NH$ ***DAA*** $X = NCH_3$ ***DAMA***

$X = \overset{+}{N}HCH_3\ CF_3COO^-$ ***DAMATFA*** $X = \overset{+}{N}(CH_3)_2\ Cl^-$ ***DADMAC***

No.	Compound	Solvent	CH_3-***	-$^1CH_2$-N(X_2-part)			$^3CH_2$= (AB-part)				-^2CH = (M- part)		
			δ	**δ_1**	**n**	**$J_{1,2}$**	**δ_{3a}**	**δ_{3b}**	**n**	**$J_{2,3a}$**	**δ_2**	**n**	**$J_{2,3b}$**
1	DAGA	D_2O	1,99	4,09	1	5,2	5,38	5,34	3	11,2	5,94	12	18,7
2	DAGA****	DMSO-d_6	1,64	3,88	2	5,4	5,18	5,19	5	10,6	5,76	12	17,2
3	DAG	$CDCl_3$	-	3,82	2	4,8	5,20	5,15	8	10,4	5,79	12	17,3
4	DAG	D_2O	-	3,88	5	4,7	5,19	5,15	10	10,3	5,79	12	17,3
5	DAGTFA	D_2O	-	4,31	2	5,4	5,62	5,57	3	13,4	6,16	10	18,7
6	DAGTFA*****	$(CD_3)_2C{=}O$	-	4,11	2	5,3	5,28	5,33	3	11,2	5,91	8	18,7
7	DAA	$CDCl_3$	-	3,21	6	6,0	5,06	5,13	16	10,3	5,86	12	17,2
8	DAMA	$(CD_3)_2C{=}O$	2,16	2,98	2	6,0	5,08	5,16	8	10,2	5,84	10	17,1
9	DAMATFA	$(CD_3)_2C{=}O$	2,79	3,79	2	7,0	5,51	5,56	3	10,2	6,07	10	17,1
10	DADMAC	D_2O	3,27	4,14	2	7,5	5,99	5,94	3	11,2	6,30	12	17,1

Notes: *The main abbreviations: **δ** – the value of chemical shift of corresponding protons, ppm; **n** – the number of lines in signal of protons of given type; **J_{ij}** – the constants of spin-spin coupling of corresponding protons, Hz;

**The number of protons by integral intensities is in agreeing with supposing structure: CH_3 – 3H (for DADMAC 6H); - CH_2 - N - 4H; CH_2 = 4H; -CH = 2H;

***Signals of methyl protons are revealed as singlets;

****Protons connected with nitrogen are revealed in given case as broad signal in the interval 6,5–11,2ppm (4H by integral intensity);

***** Protons connected with nitrogen are revealed in given case as broad signal with the center at 7,96ppm (4H by integral intensity).

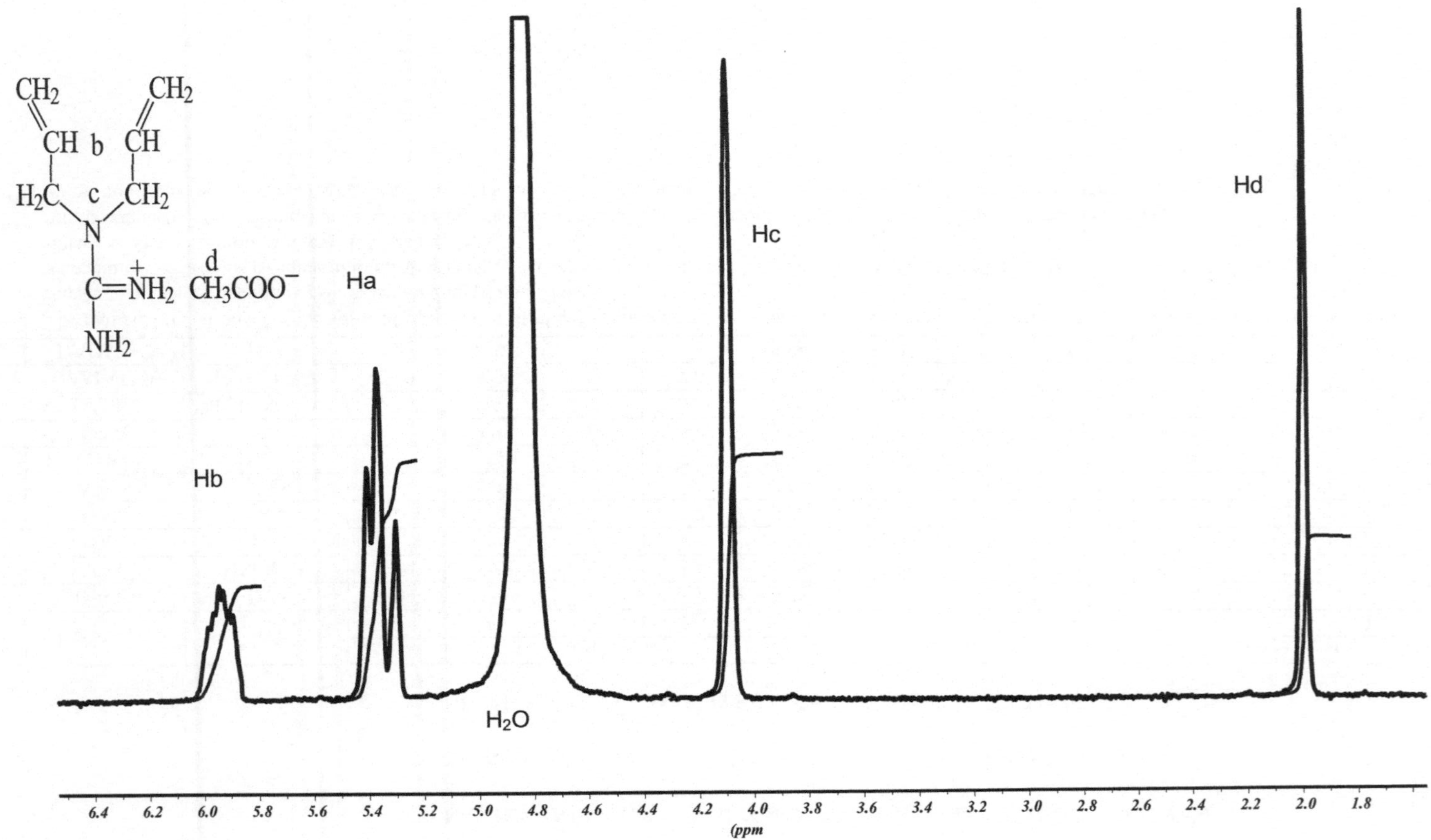

Figure 4.6. ^{1}H NMR spectrum of diallylguanidine acetate in D_2O.

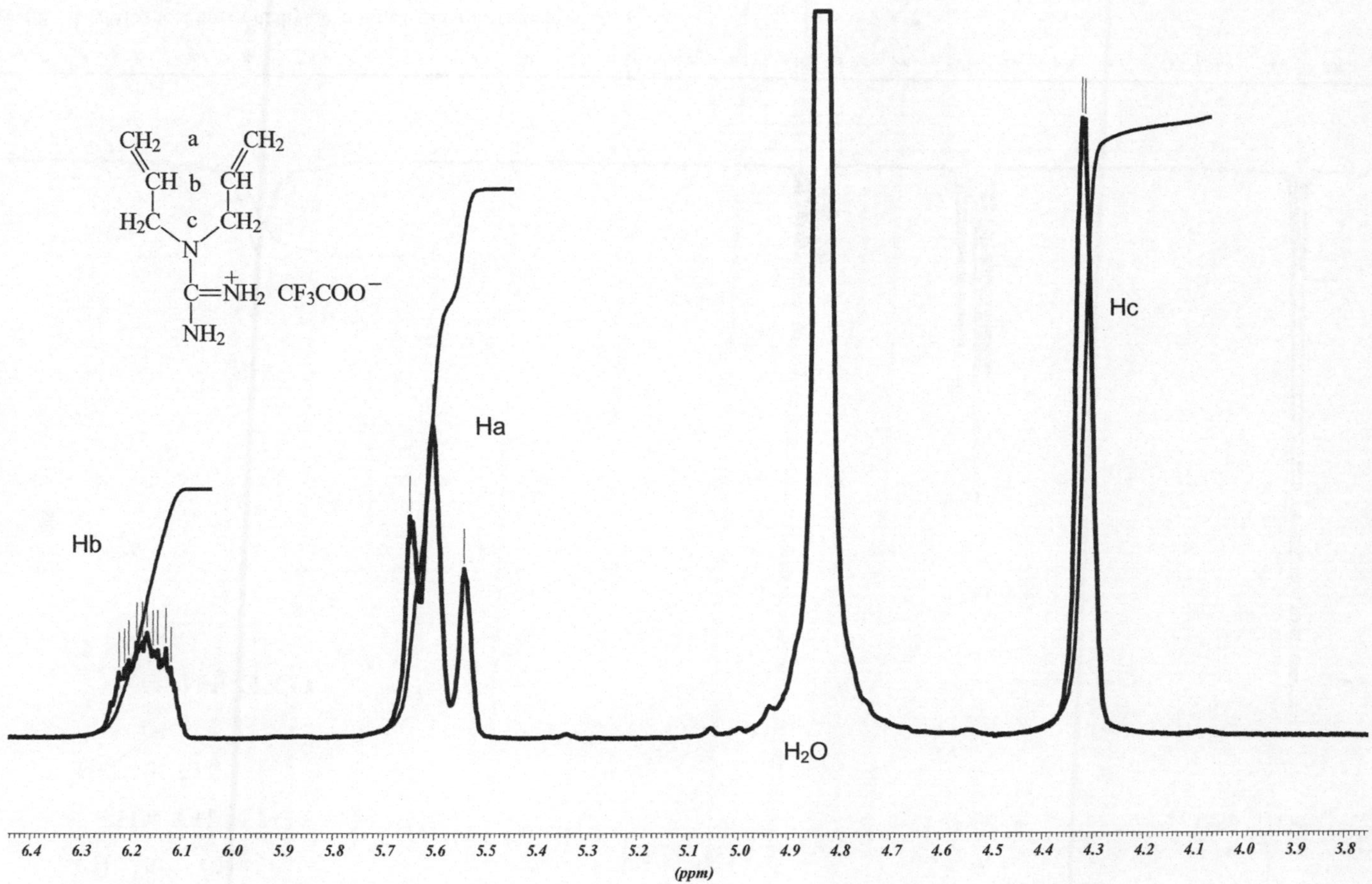

Figure 4.7. ^{1}H NMR spectrum of diallylguanidine trifluoroacetate in D_2O.

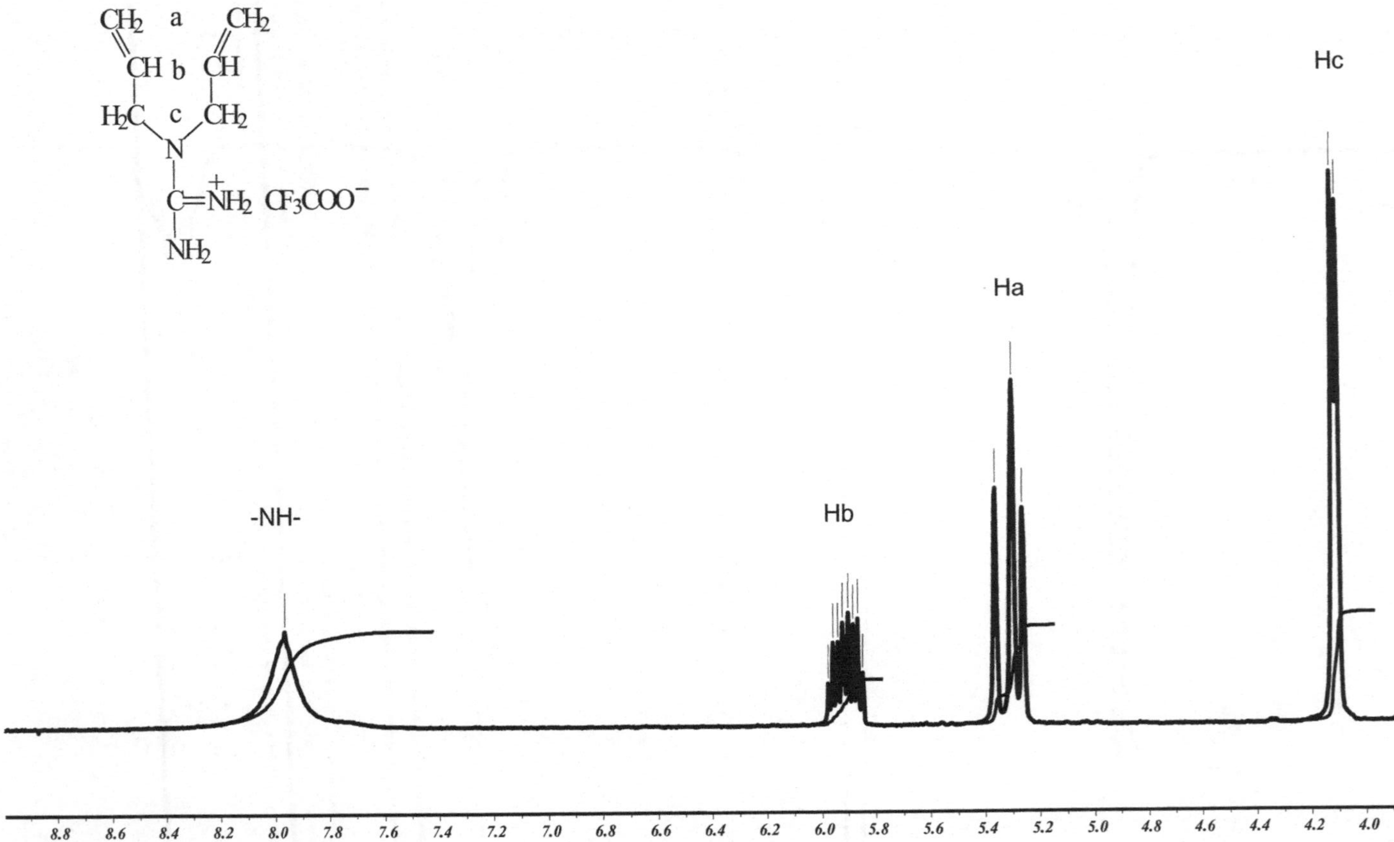

Figure 4.8. ^{1}H NMR spectrum of diallylguanidine trifluoroacetate in $(CD_3)_2C{=}O$.

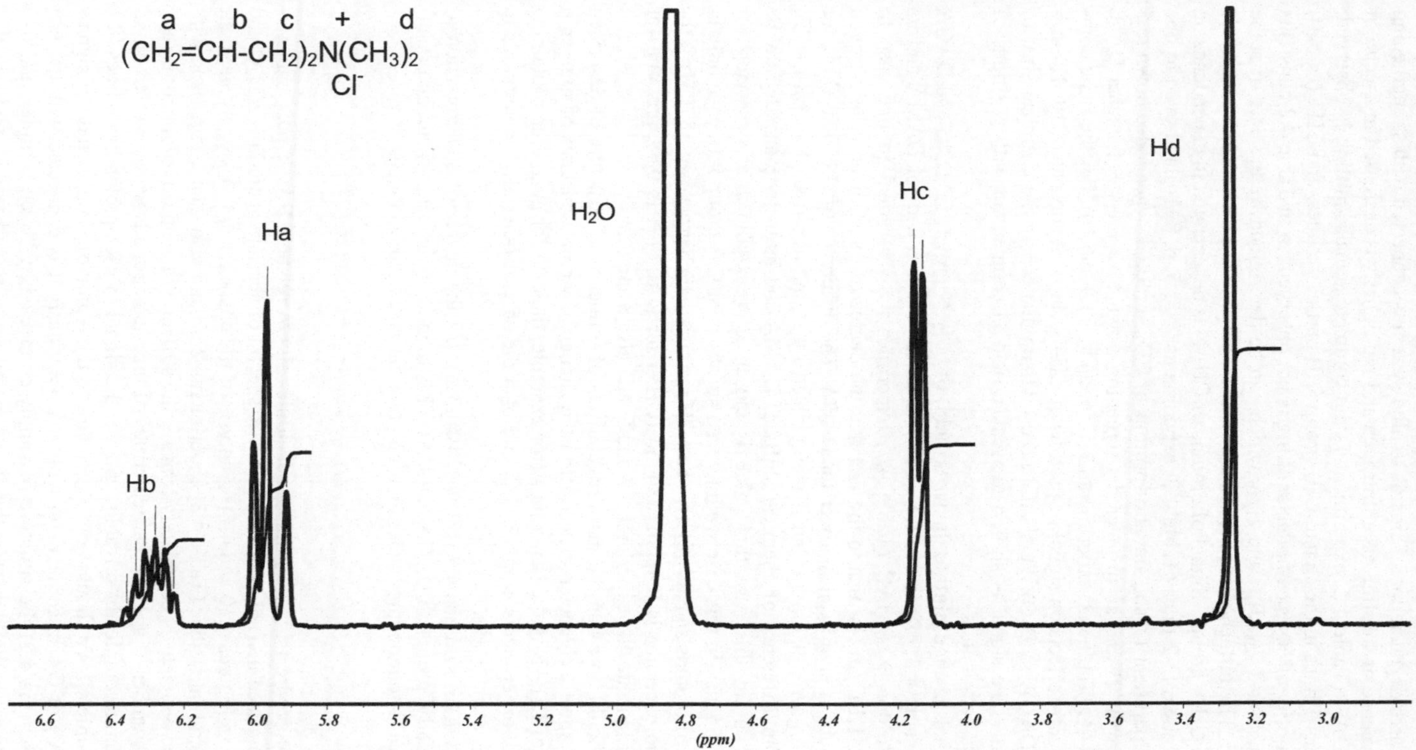

Figure 4.9. ^{1}H NMR spectrum of N,N-diallyl-N,N-dimethylammonium chloride in D_2O.

In the majority of cases degeneration of signals of allyl system is connected with strengthening of dis-screening at the expense of substituents at nitrogen and multiplicity is decreased in spectra of $ABMX_2$ type. Only for DAA (Table 4.2, experiment 7) the complete signals splitting is observed: CH_2-N group (X_2 part) has 6 lines in the signal, CH_2-group (AB part) – 16 lines, and CH= (M part) 12 lines. Since in the majority of cases complete splitting is not observed, so in Table 4.2 constants of spin–spin coupling such as $J_{3a,3B}$, $J_{1,3a}$ and $J_{1,\ 3b}$ which can be determined by the data of some spectra are not presented. The influence mentioned above is revealed to a greater extent on signals of protons of CH_2-N and CH_2= groups. Signal of protons of CH_2-N is degenerated in dependence on substituents and solvent down to 2 lines (Table 4.2, experiments 2, 3, 5, 6, 8–10) and even to one for DAGA (Table 4.2, experiment 1), and signal of protons of CH_2= group is degenerated down to 5 lines (Table 4.2, experiment 2) and at stronger dis-screening down to 3 lines – for slats DAG, DAMATFA and DADMAC (Table 4.2, experiments 1, 5, 6, and also 9 and 10). For the same compounds this influence slightly effects on signal of protons of CH= group – it consists of 10–12 lines (with one exception of DAGA in D_2O).We should note that simultaneously the chemical shift of these protons is changed – the signal is shifted to weaker field in comparison with positions of analogous signals of other compounds.

Comparison of signals change (shift to the weak field) of CH_2–N groups for DAG and DAGA caused by protonation (Table 4.2, experiments 1 and 4) equal to 0,21ppm and for DAMA and its trifluoroacetate, and also for DADMAC (Table 4.2, experiments 8–10) equal in one case to 0,81ppm and in other to 1,16ppm testifies to the fact that the degree of quaternization of nitrogen connected with allyl groups is significantly lower for DAGA. The difference of chemical shifts for DAG and DAGTFA is higher – 0,43ppm, but it is explained by strong negative –I–effect of CF_3 group of counter-ion. The difference of chemical shifts of DAMA and its trifluoroacetate may be explained analogously, but in this case this value is closer to weak-pole shift observing for DADMAC (0,81 and 1,16ppm). In this case we may also speak about CF_3 group influence only in comparison of positions of protons signals of CH_2= and CH= groups which are close: $\Delta\delta$ (CH_2=) for DAG and DAMA trifluoroacetates is 0,4ppm, and $\Delta\delta$ (CH=) for DAGTFA is 0,37ppm, and for DAMATFA it is 0,23ppm (Table 4.2, experiments 4 and 5, and also 8 and 9).

Additionally we should note that the main regularities of influence of structure of described compounds on ^{1}H NMR spectra characteristics confirm the change of vicinal constants of spin–spin coupling $J_{1,2}$ of protons of CH_2–N and CH= groups (for example, the data for guanidine containing compounds in Table 4.2, experiments 4 and 2, 5, 6 and also the data of other allyl compounds in Table 4.2, experiments 7, 8 and 9, 10).

It is consequent from all said above (as it was assumed) that the degree of quaternization of nitrogen connected with allyl groups in DAGA and DAGTFA monomers is insignificant and insufficient for inhibition of DCT to monomer under their polymerization by radical mechanism.

4.2.2. Vinyl derivatives

^{1}H NMR spectra of acrylic acid (AA) and its guanidine salt GA relate to BAC type. Characteristic of signals are summarized in Table 4.3 and ^{1}H NMR spectra of acrylate guanidine in D_2O and DMSO-d_6 are presented in Figures 4.10 and 4.11. Analysis of the character of GA signals and their chemical shifts and integral intensities (Table 4.3, experiments 2, 3) and also their comparison with data on AA (Table 4.3, experiment 1) completely confirm the structure of synthesized monomer salt – acrylate guanidine. A number of spectral particularities of these compounds were revealed. In all cases ^{1}H NMR spectra have completely solvable type, i.e. the signal of each proton has 4 lines (the so-called doublets of doublets). We should note slight shift to the region of stronger field of signals of methylene protons (^{3}C) of GA in comparison with AA. Obviously it is connected with the fact that for GA in water (Scheme 4.2) the structure of simply connected hydrogen complex and (or) monomer is more characteristic that decreases the dis-screening action of carboxyl group to insufficient degree.

Table 4.3. Spectral characteristics of acrylate derivatives*, **.

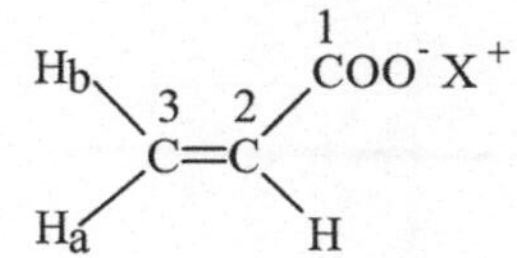

X = H Acrylic acid
X = $H_2N^+ = C(NH_2)_2$ acrylate guanidine

Compound	Solvent	^3H_A				^3H_B			^2H		NH
		δ_{3a}	n	$J_{3a,2}$	$J_{3a,3b}$	δ_{36}	n	$J_{3a,2}$	δ_2	n	δ
AA	D_2O	5,94	4	9,63	1,05	6,36	4	17,13	6,11	4	-
GA	D_2O	5,91	4	9,64	2,14	6,27	4	17,67	6,41	4	-
GA	DMSO-d_6	5,28	4	9,64	3,21	5,79	4	17,14	5,96	4	7,74

Notes: *The main abbreviations: **δ** – the value of chemical shift of corresponding protons, ppm; **n** – the number of lines in signal of protons of given type; J_{ij} – the constants of spin-spin coupling of corresponding protons, Hz;

**The number of protons by integral intensities is in agreeing with supposing structure: 1H for all protons of vinyl system and 6H for guanidine counter-ion (it is revealed as broaden singlet).

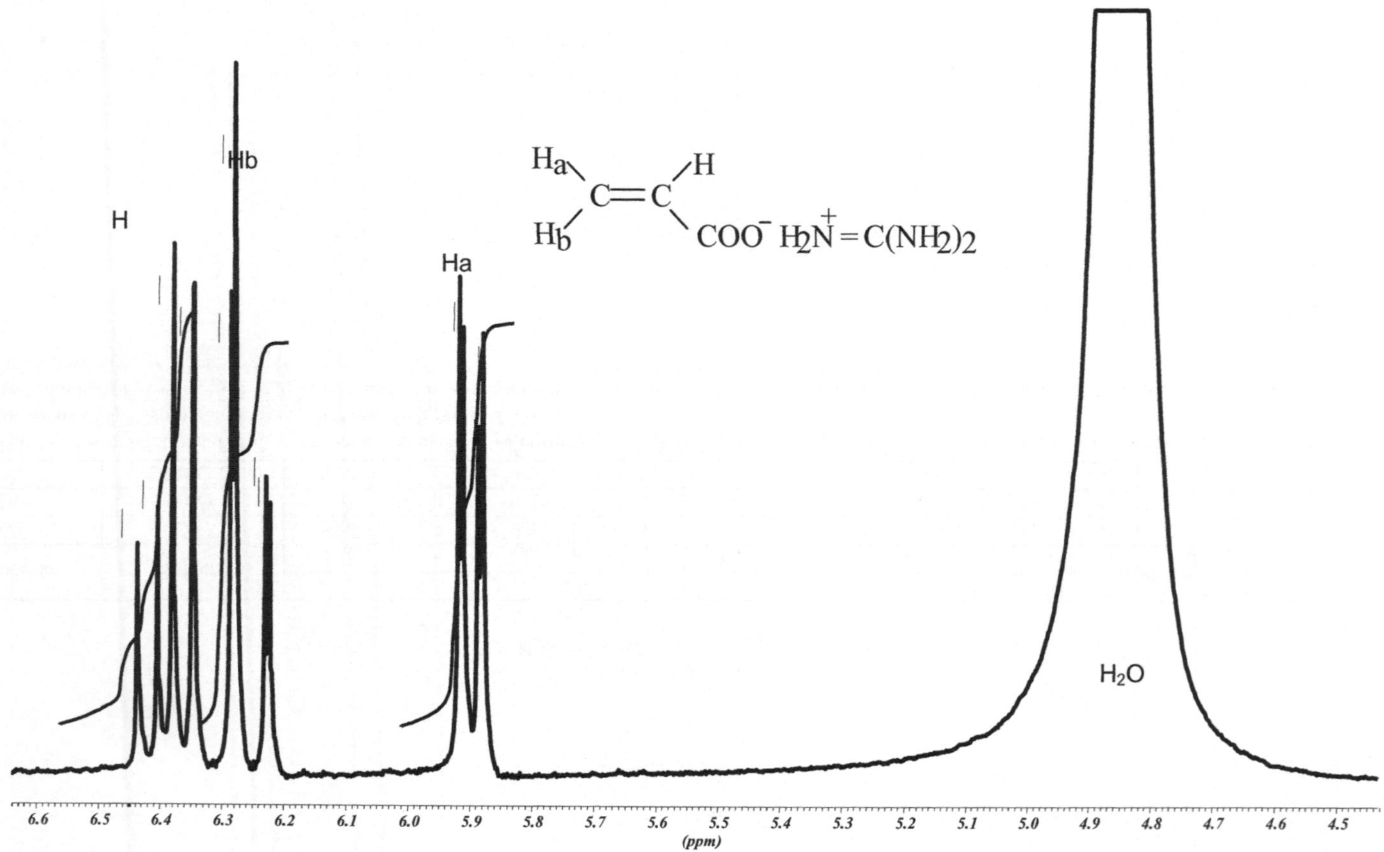

Figure 4.10. ^{1}H NMR spectrum of acrylate guanidine in D_2O.

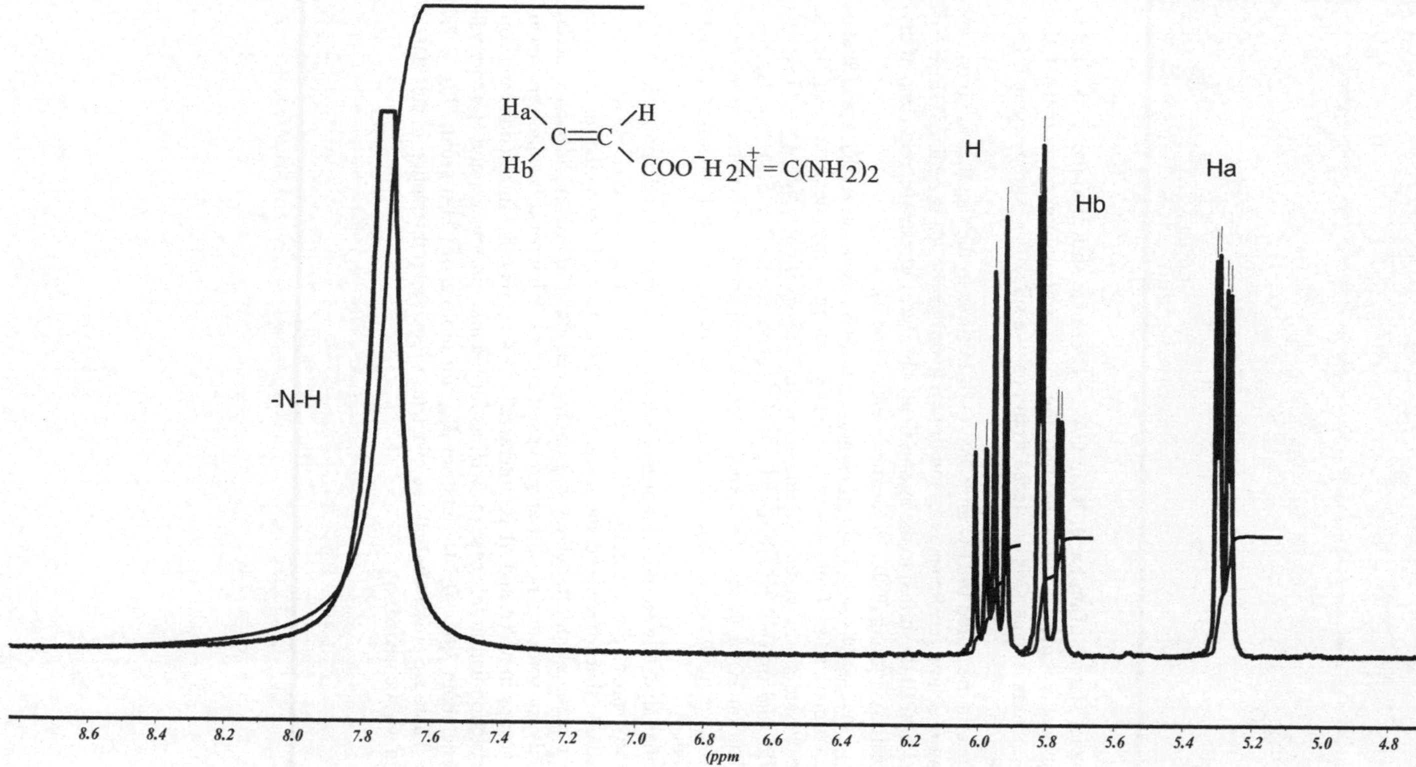

Figure 4.11. ^{1}H NMR spectrum of acrylate guanidine in DMSO-d_6.

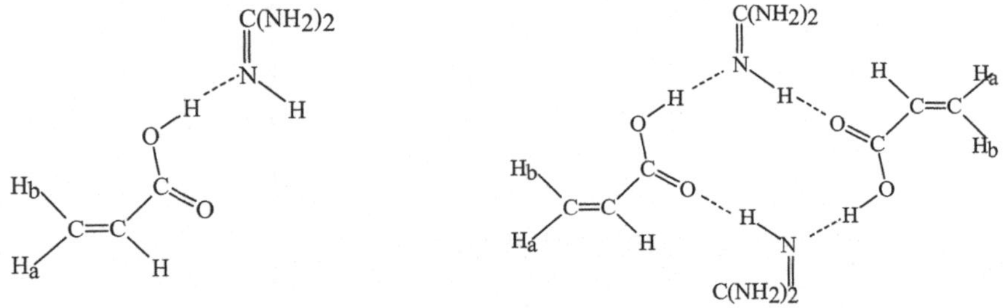

Scheme 4.2.

From the other hand, signals of proton at ^2C in GA spectrum are shifted to the weak field in comparison with AA; probably it may be connected with the change of GA conformation in solution in comparison with AA, and proton at ^2C is displaced from positive side of anisotropy cone of C=O group to the negative region.

Comparison of spectra measured in D_2O and DMSO-d_6 testifies to the fact that in given case the first type of bonding (Section 4.1, Scheme 4.1) with two hydrogen bonds is realized, since signals of all protons are significantly shifted to the strong field. This fact indicates on reduction of dis-screening influence of carboxyl group (Table 4.3, experiments 2 and 3).

^1H NMR spectra of methacrylic acid and its guanidine salt MAG relate to ABX_3 type. Characteristics of signals are summarized in Table 4.4 and ^1H NMR spectra of methacrylate guanidine in D_2O and DMSO-d_6 are presented in Figures 4.12 and 4.13. the analysis if character of MAG signals and their chemical shifts and integral intensities (Table 4.4, experiments 2 and 3) and also their comparison with the data on MAA (Table 4.4, experiment 1) completely confirm the structure of synthesized monomer sal;t – methacrylate guanidine. a number of spectral particularities of these compounds was revealed.

We should note that signals were not completely solvable, i.e. there was degenerated ABX_3 type of spectra. It may be connected with strong influence of CCOX group (especially in the case of MAG). We couldn't exclude that for given compounds as in the case with some other diallyl compounds (Table 4.2) the apparatus work frequency was not high enough (300MHz) and such structure of lines was revealed only for some but not for all groups of signals. It happened so since protons of methyl and methylene groups in MAA and MAG interacted via 4 bonds (allyl interaction) and constants of spin-spun interaction had in this case (as well as for heminal protons H_A and H_B) low values (for the $^4\mathbf{J}$ magnitude about 3,0 – -0,5 Hz, and for $\mathbf{J_{hem}}$ for protons of CH_2= group -3,0 – +2,0 Hz). That is why in the Table we present only those constants of spin-spin interaction in calculation of which by spectra data we succeeded.

Table 4.4. Spectral characteristics of methacrylate derivatives*, **.

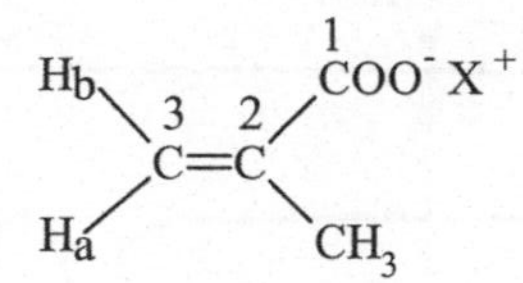

X = H Methacrylic acid

$X = H_2N^+ = C(NH_2)_2$ Methacrylate guanidine

Compound	Solvent	CH_3				^3H_A			^3H_B		NH
		δ	**n**	$J_{Me,3a}$	$J_{Me,3b}$	δ_{3a}	**n**	$J_{3a,Me}$	δ_{3b}	**n**	**δ**
MAA	D_2O	2,10	3	1,62	1,05	5,90	3	1,62	6,31	1	-
MAG	D_2O	2,08	3	2,13	1,08	5,53	1	-	5,86	1	-
MAG	DMSO-d_6	1,77	1	-	-	5,04	3	1,62	5,60	2	7,67

Notes: *The main abbreviations: **δ** – the value of chemical shift of corresponding protons, ppm; **n** – the number of lines in signal of protons of given type; J_{ij} – the constants of spin-spin coupling of corresponding protons, Hz;

**The number of protons by integral intensities is in agreeing with supposing structure: 1H for methyl protons and 3H for methyl protons and 6H for guanidine counter-ion (it is revealed as broaden singlet).

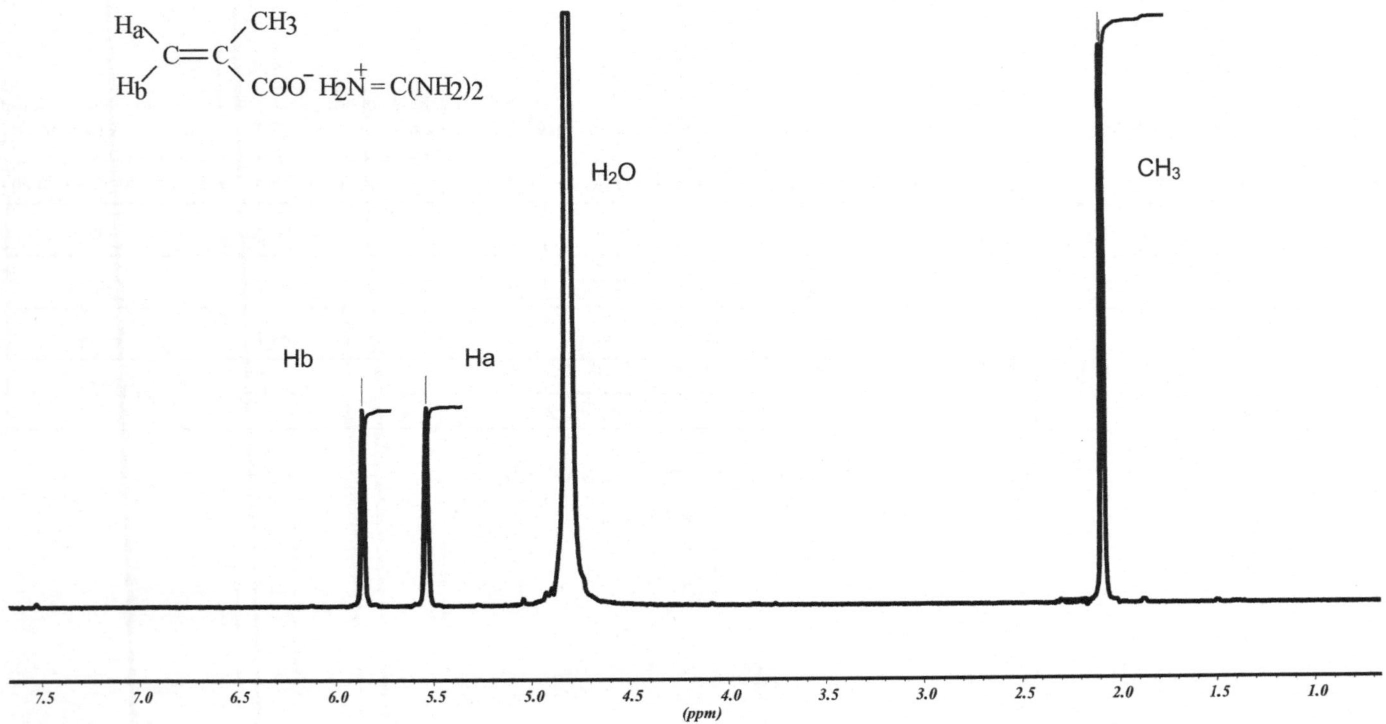

Figure 4.12. ^1H NMR spectrum of methacrylate guanidine in D_2O.

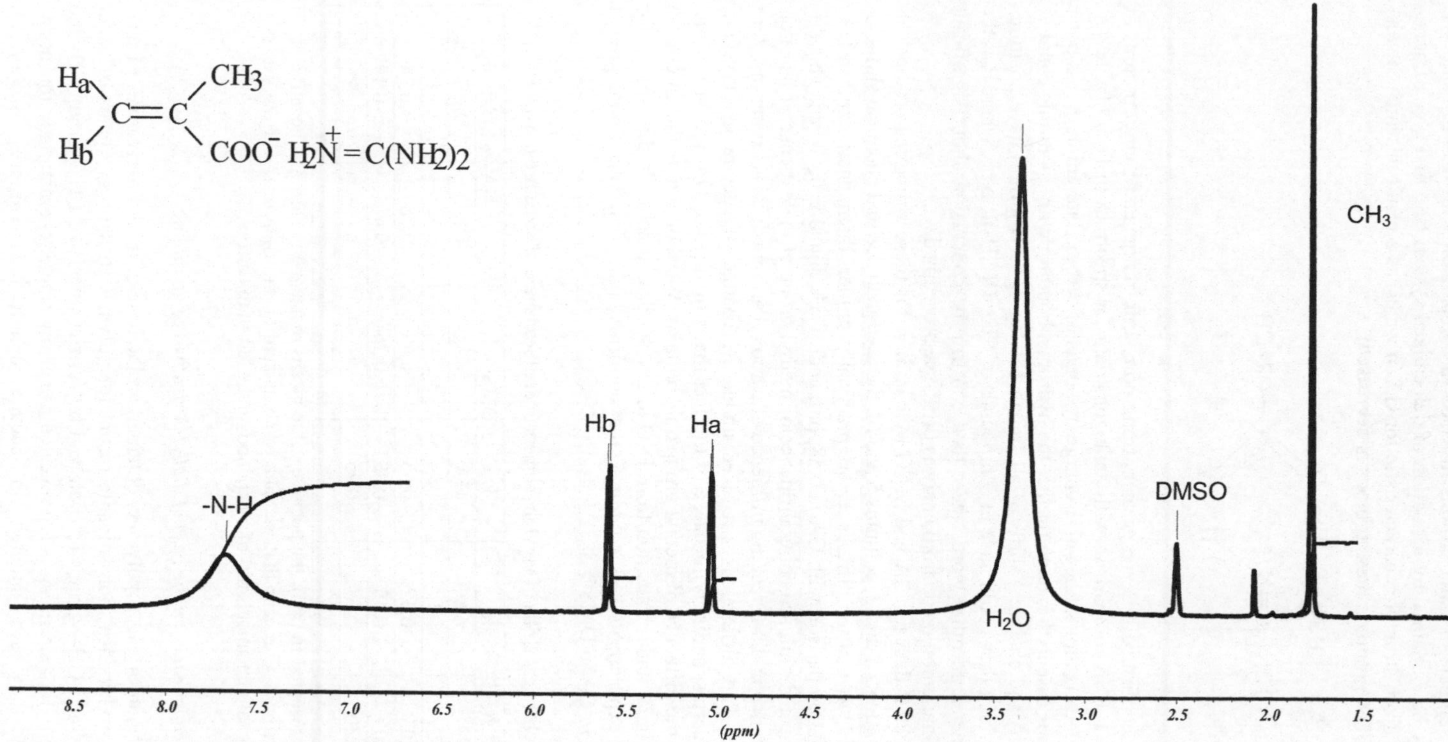

Figure 4.13. ^{1}H NMR spectrum of methacrylate guanidine in DMSO-d$_6$.

Strong pole shift of signals of vinyl protons in MAG in comparison with MAA (Table 4.4, experiments 2 and 1) may be explained by formation of delocalized system (the first type of bonding, scheme 4.1, differing from structure characteristic for GA in water) as a result of which the dis-screening influence of COOX group on these protons is decreased:

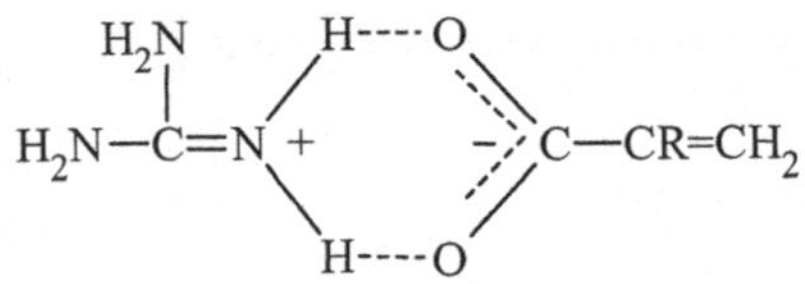

This supposition is confirmed by even greater strong pole shift of signals of vinyl protons of MAG measured in DMSO-d_6 in comparison with analogous data for spectrum in D_2O (Table 4.4, experiments 3 and 2). Delocalization of negative charge by bonds of carboxylate-anion in less polar than water DMSO becomes essential enough (at that the hydrogen bonding with guanidine counter-ion is probably strengthen) and chemical shifts of all signals (including CH_3-group) have positions very close to those ones which they could have in compounds of $CH_2{=}C(CH_3)R$ type, where R is the substituent of alkyl character. In other words these data confirm the conclusion about significant delocalization in carboxylate-anion made from analysis of IR-spectroscopy data.

Of we compare spectral data for GA and MAG by common for these structures protons ^3H_a and ^3H_b (Table 4.5, experiments 1 and 2, columns 5 and 6), then we shall see that chemical shifts for MAG are in stronger field. At that the differences are practically similar: diamagnetic shift in D_2O of ^3H_A signal for MAG in comparison with GA is 0,38ppm, and of ^3H_B signal is 0,41ppm. This difference obviously is connected with screening influence of methyl group at the expense of its +I effect and also due to difference in dis-screening influence of carboxyl group which is higher in GA as it was mentioned above. The difference is reduced under the change of spectra in DMSO-d_6 (0,24ppm and 0,19ppm, correspondingly) because of the fact that structures of GA and MAG salts in this case are equal, so, the influence of methyl group only remains. We should note that in GA the diamagnetic shift of ^2H signal under transition from D_2O to DMSO-d_6 is practically the same as for other protons (Table 4.5, experiments 1–3, column 3) that confirms higher delocalization degree of charge of carboxyl group in GA in DMSO-d_6.

Table 4.5. The change of chemical shifts of vinyl monomers under the change of structure or solvent*, **.

No.	Protons	Acid → Salt		D_2O → DMSO-d_6		GA → MAG	
		GA	MAG	GA	MAG	D_2O	DMSO-d_6
	(columns)	1	2	3	4	5	6
1	^3H_A	0,03	0,37	0,63	0,49	0,38	0,24
2	^3H_B	0,09	0,45	0,48	0,26	0,41	0,19
3	^2H	-0,30	-	0,45	-	-	-
4	CH_3	-	0,02	-	0,31	-	-

Notes: * – All values are presented in ppm; the positive value means diamagnetic shift and negative one – paramagnetic. ** – The arrow "→" means that the data of compound on the right of the arrow was subtracted from the value of signal's chemical shift of compound on the left of the arrow.

4.2.3. Homopolymers and copolymers on the base of diallyl derivatives of guanidine

^1H NMR spectra of homo- and copolymers obtained on the base of diallyl derivatives of guanidine are characterized by broaden, non-solvable (usual for polymer structures) signals of chain CH_2– groups and CH_2–N and CH– groups of cycle and by narrow signals of CH_3– groups. By the way methyl signals of diallylguanidine units are revealed as narrow singlet characteristic for monomer compounds since they are situated in acetate counter-ions (Figures 4.14–4.17).

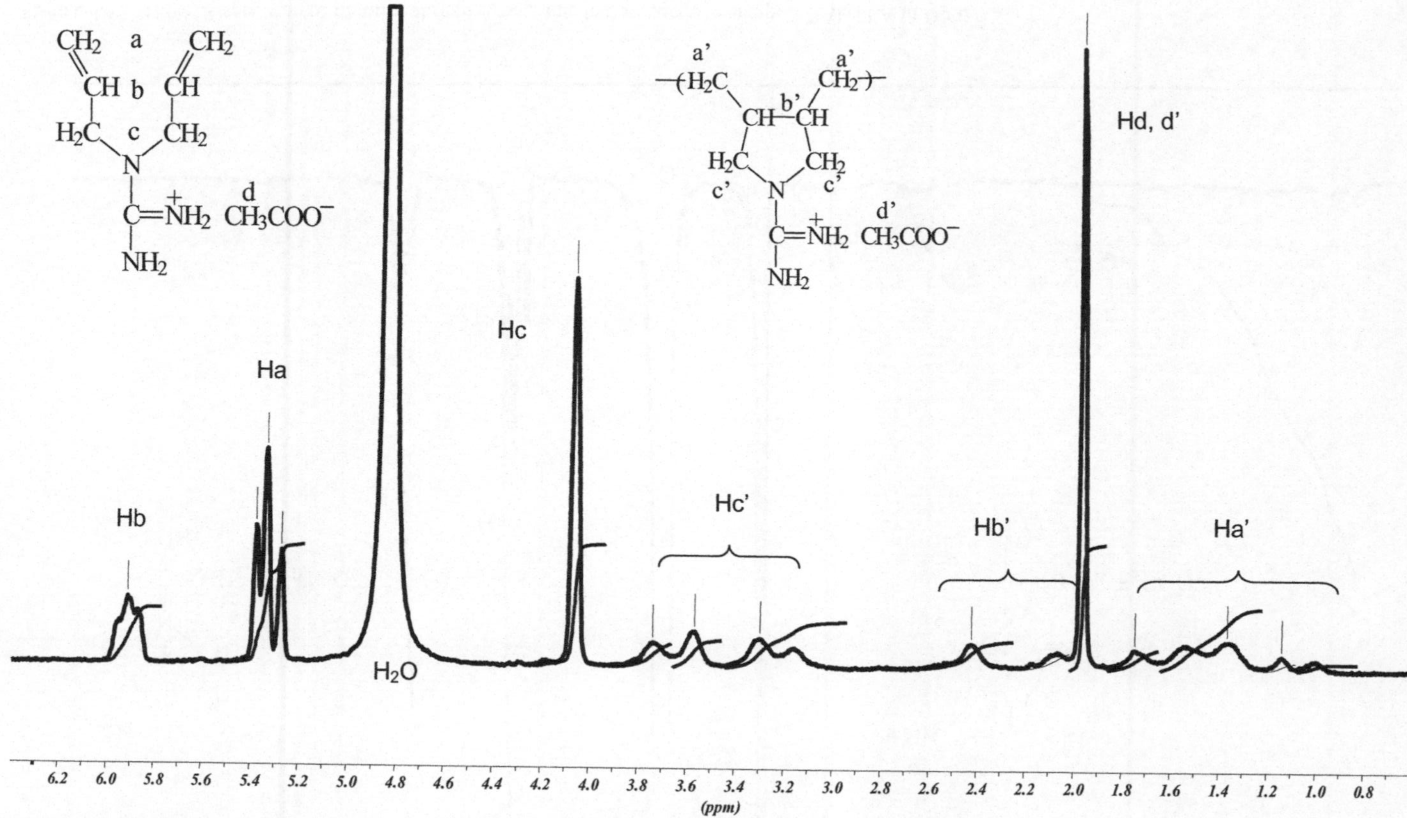

Figure 4.14. ^{1}H NMR spectrum of mixture of diallylguanidine acetate and polyDAGA in D_2O.

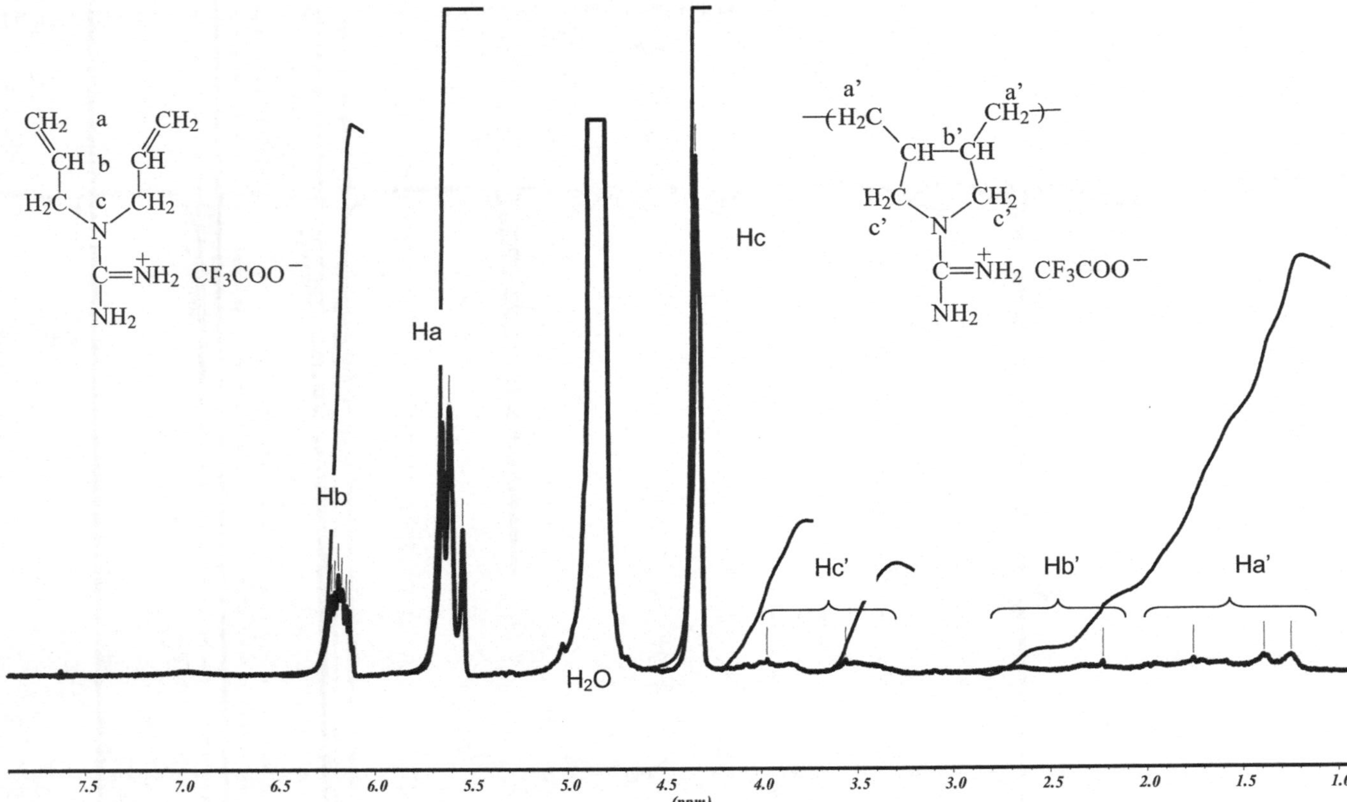

Figure 4.15. ^{1}H NMR spectrum of mixture of diallylguanidine trifluoroacetate and polyDAGTFA in D_2O.

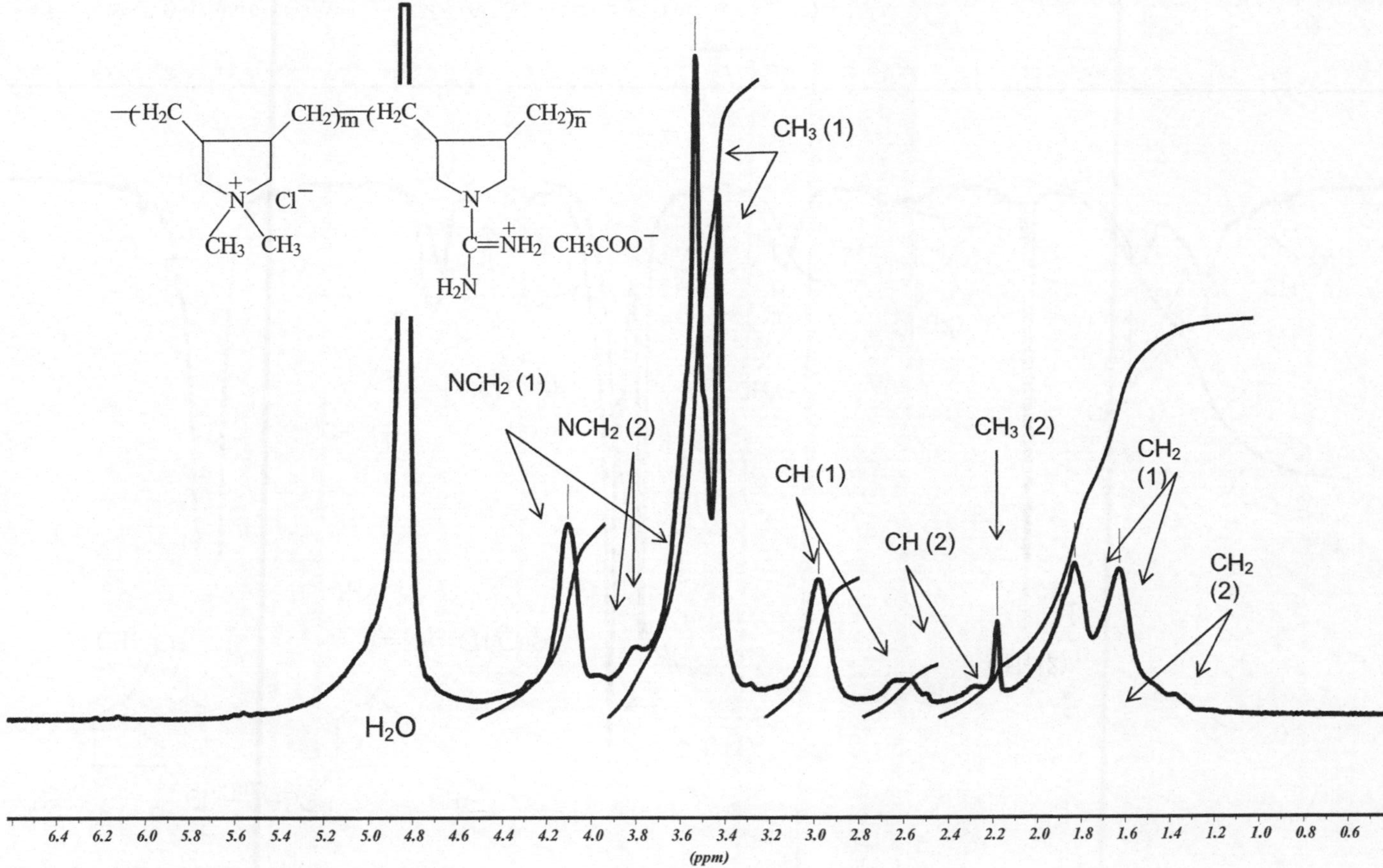

Figure 4.16. ^{1}H NMR spectrum of copolymer of DADMAC (**1**) and DAGA (**2**) (m : n = 87 : 13) in D_2O.

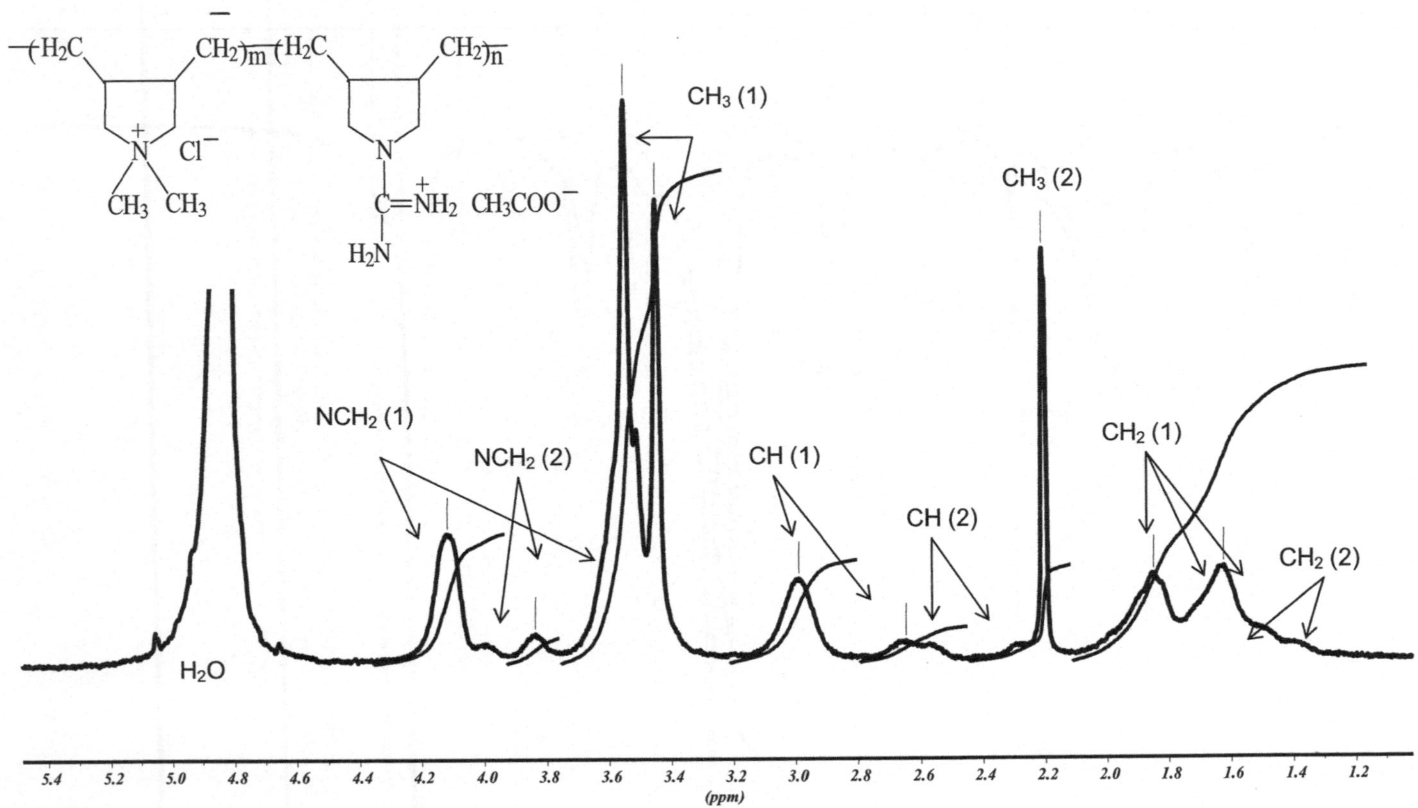

Figure 4.17. ^{1}H NMR spectrum of copolymer of DADMAC (**1**) and DAGA (**2**) (m : n = 75 : 25) in D_2O.

Homopolymers DAGA and DAGTFA (Figures 4.14, 4.15) are characterized by four broaden bands in the region of 1,5 ppm of protons of CH_2-groups of the chain, by 4 the similar signals in the region of 3,5 ppm of CH_2–N group of cycle, and also by 2 bands in the region of 2,4 ppm of CH-group of cycle. At that signals of PDAGA are situated in stronger filed in comparison with signals of PDAGTFA that is explained by strong negative induction effect of trifluoromethyl group of counter-ion.

^{1}H NMR spectrum of PDADMAC (Figure 4.18) is characterized by broaden non-solvable (usual for polymer structures) signals of CH_2– groups (4 lines) of chain, of CH_2–N (2 lines) and CH– groups (2 lines) of cycle and narrow signals of CH_3– groups (2 lines). Signals of PDADMAC in comparison with corresponding monomers have diamagnetic shift of protons of CH and CH_2 groups (corresponding to monomer CH_2= and CH= groups) connected with formation in structure of substituents of alkyl character after polymerization, whereas signals of methyl protons have paramagnetic shift due to transformation under polymerization into alkyl electron-acceptor vinyl groups decreasing dis-screening action of positive charge at nitrogen. For protons of CH_2N groups one of two signals is revealed at the same place where monomer's signal is, and the second one is shifted to stronger field and is covered by signals of methyl groups. At that ratio of signals of given protons is ~ 1:1, methyl groups - ~ 2:1 (more intensive signal is revealed in weaker field), signals of CH groups – ~ 5:1 (the signal larger by integral is in weaker field) and for four signals of CH_2 groups of chain the ratio is the same is one compare more intensive central signals with weak neighboring signals. Formation of a large number of lines in spectra of homopolymers and DADMAC copolymers (in comparison with monomer) for corresponding groups of protons is connected first of all with appearance under polymerization of cyclic structures that in combination with polymer character of compounds leads to such result. That is why methyl protons of DADMAC units for example are revealed in spectrum in the shape of two signals, and protons of CH_3 group of DAGA unit in the shape of one signal, since they are situated in counter-ion and are not included into carbo-chain structure of polymer. Due to the same reason nonequivalence of protons of CH_2 groups which is observed for protons of monomer CH_2= groups remains.

The larger number of signals for protons of CH_2–N groups in PDAGA and PDAGTFA in comparison with PDADMAC is probably connected with difference in spatial structure of these homopolymers. In the case of diallylguanidine polymers the formation of such system of hydrogen bonding when one counter-ion is connected with two units of monomer included into the chain is possible that may cause significant nonequivalence of above mentioned protons as a result of peculiar fixation of spatial structure.

Obviously, this fact is connected with the change of structure IV characteristic for DAGA monomer by structure V (Scheme 2.5, section 2.1.2) revealing in polymer that is caused by the fact that structure V is less spatially hindered and more convenient for formation of system of hydrogen bonding by polymer structure. The indirect confirmation of this supposition is the fact that in copolymer DAGA and DADMAC the signals of protons of CH_2–N group have two bands as in PDADMAC that is connected by the fact that DAGA units are separated in the chain by DADMAC units.

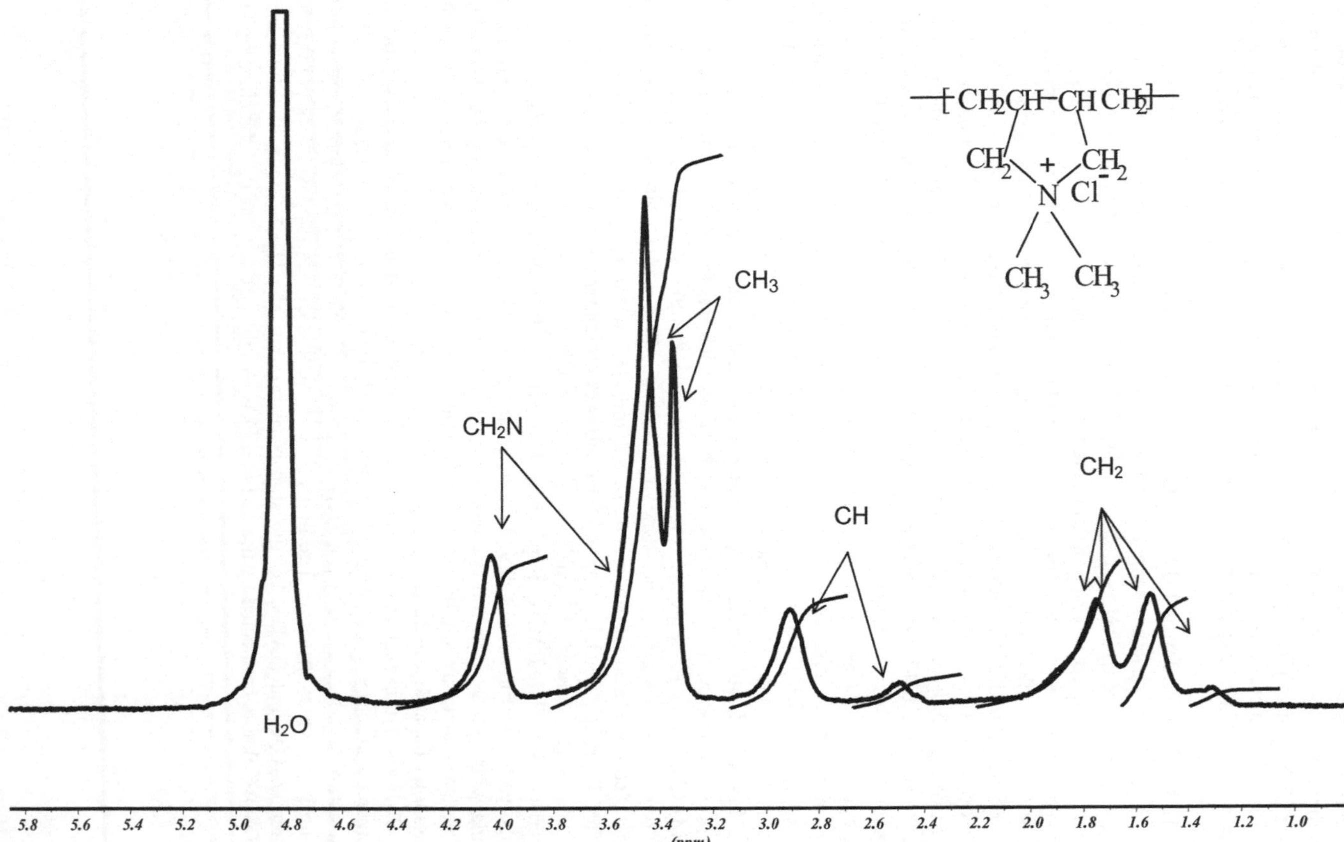

Figure 4.18. ^{1}H NMR spectrum of polydiallyldimethylammonium chloride in D_2O.

In copolymers of DADMAC with DAGA (Figures 4.16 and 4.17) the signals of protons of chain and cycle may be easily related to various comonomers (with consideration of various integral intensities). However many of them are covered, so under analysis of copolymer structure the integral intensities of methyl protons were considered first of all. We should note that the majority of copolymers (as well as those ones presented in Figures 4.16 and 4.17) is enriched by DADMAC units and DAGA units don't influence on signals of DADMAC units. However the dis-screening influence of DADMAC units possessing positive charge in the main carbo-cyclic chain leads to the shift of all signals of DAGA units to the weak filed (Figures 4.16, 4.17 and 4.14). And this shift is more significant than analogous change of chemical shifts of protons in PDAGTFA in comparison with PDAGA that is connected with the influence of CF_3 group of counter-ion. This fact, firstly indicates on lower degree of nitrogen quaternization in the structure of diallylguanidine derivatives in comparison with DADMAC, and secondly it testifies that DAGA units are situated in compact structure and spatial surrounding of DADMAC units since the dis-screening influence through the space corresponds to that one transmitted by bonds. Chemical shifts of protons of DAGA units become close in their values to chemical shifts on protons of DADMAC units.

In conclusion of this Section we suggest the order of calculation of comonomer structure of copolymer of DADMAC with DAGA on the base of consideration of integral intensities of various signals (see Figures 4.16 and 4.17). since the signal of methyl group of DAGA is not covered by other signals then the one third of its integral intensity will be equal to the value of conventional proton for the unit DAGA–"$1H_{PDAGA}$". The rest protons relating to signals of chain and DAGA cycle and also to signals of methyl groups of DADMAC chain and cycle are covered that is why for determination of conventional proton of DADMAC unit "$1H_{PDADMAC}$" out of total integral intensity (**I**) one should subtract the contribution of ten protons of chain and cycle of DAGA unit and obtained value should be divided by 16 protons of DADMAC unit (equation 5):

$$"1H_{PDADMAC}" = (I - 10 \times "1H_{PDAGA}") : 16 \qquad (5)$$

From obtained results one may easily determine the molar content of comonomers in copolymer (**M**$_{DADMAC}$ and **M**$_{DAGA}$) expressed in mol/ % (equations 6 and 7):

$$M_{DADMAC} = ["1H_{PDADMAC}" : ("1H_{PDADMAC}" + "1H_{PDAGA}")]\cdot 100\% \qquad (6)$$

$$M_{PDAGA} = ["1H_{PDAGA}" / ("1H_{PDADMAC}" + "1H_{PDAGA}")]\cdot 100\% \qquad (7)$$

4.2.4. Homopolymers of (meth)acrylateguanidines and their copolymers with diallyldimethylammonium chloride

4.2.4.1. Homopolymers

The ordinary number of broaden bands usual for polymers is observed in ^1H NMR spectra (Table 4.6) of polyacrylate guanidine (PAG) (Figure 4.19) and polymethacrylate guanidine (PMAG) (Figure 4.20). As it was mentioned above for PDADMAC, the signals of homopolymers in comparison with corresponding monomers have diamagnetic shift of protons of CH and CH_2 groups (corresponding to monomer CH_2= and CH= groups) connected with formation of aliphatic structure of polymer chain after polymerization. Due to the same reason the signals of methyl protons in PMAG spectrum also have diamagnetic shift since after polymerization dis-screening action of double bond connected with methyl group disappears. In PAG spectrum two lines relate to protons of CH_2 the most intensive of which is revealed in stronger field at 1,68ppm (integrals ratio is ~ 4:1) and one broad band of signals of protons of CH reveling in weaker field in comparison with methyl protons. It is a consequence of neighboring carboxyl group influence. In PMAG spectrum CH_3 substituent is revealed as 4 lines reducing in intensity as far as paramagnetic shift is increased and the most intensive line is situated in strong field and its ratio to the sum of the rest ones is ~ 7:3. Protons of CH_2 groups are revealed in the shape of two lines and as well as for methyl groups the most intensive band is situated in strong field and the signals ratio is ~ 4:1. The presence of such number of bands is connected with the fact that polymerization of given monomers may proceed as by "head–to–tail" so by "head–to–head" type. When the structure is formed as a result of "head–to–head" interaction in spectrum the signals relating to protons of methyl and methylene groups should be revealed in stronger field since in this case the dis-screening influence of carboxyl groups is lower. In the case when structure is formed as a result of "head–to–tail" interaction the signals of protons of methyl and methylene groups should be revealed in weaker field. In addition to these facts we should note that analysis of ball-rod models with various lengths of bonds allowed determining the following particularities of GA and MAG polymerization processes and formation of structure of resulted polymers. Under the attack "head–to–head" the hydrogen bonding of two carboxyl groups with one guanidine counter-ion is formed which leads to advantage spatial disposition of end group of propagating radical and monomer molecule in regular act of chain propagation. As a result such structure of polymer chain is formed which is stabilized by the system of hydrogen bonds, moreover in the case of MAG it is additionally stabilized by hydrophobic interaction of methyl substituents.

We should note that ^{1}H NMR spectra of PAA and PMAA for which formation of such stabilizing structures during radical polymerization is impossible have essential differences from PAG and PMAG spectra (Table 4.6, experiments 1, 4 and 2, 3, 5, 6). So, in PAA spectrum the protons of corresponding groups are revealed in the shape of two more broad signals than in PAG spectrum, and these two signals are partially covered. In PMAA spectrum signals of CH_2 groups are revealed as one broad singlet and methyl group gives three signals, the ratio if intensities if these signals is ~ 1:2:1. (The analogous picture is observed for polymethylmethacrylate in which spectrum there is a large number of methyl signals, especially of methylene protons – 10 lines, that indicates on low stereoregularity of given polymer.)

The additional confirmation of differences in the type of hydrogen bonding which we observed by data of ^{1}H NMR spectra for GA and MAG was obtained while comparing spectral data of corresponding guanidine containing polymers and polymer acids (Table 4.6, experiments 1–3 and 4–6). Signals of PMAG in comparison with PMAA undergo diamagnetic shift (as in the case of monomer compounds, Table 4.5) that is connected with maintenance of the first type of bonding (Scheme 4.1) with participation of delocalized carboxyl anion that decreases its dis-screening influence. Analogous comparison of PAG and PAA testifies to the fact that in this case the suggested for GA monomer structures (Scheme 4.2) also remain, since in the PAG case we observe paramagnetic (but not weak diamagnetic as for monomer compounds, Table 4.5) shift of signals of corresponding protons that is connected with polymer nature of compound: the number of neighboring carboxylate-anions (in dependence on conformation) reveal dis-screening influence on signals of CH_2 groups.

Table 4.6. ^{1}H NMR spectral characteristics of acrylate homopolymers*.

No.	Compound	CH_3				CH_2		CH
	lines**	1	2	3	4	5 (1)**	6 (2)**	(3)**
1	PAA	-	-	-	-	1,28	-	2,20
2	PAG	-	-	-	-	1,68	1,82	2,28
3	PAG	-	-	-	-	1,69	1,82	2,29
4	PMAA	1,71	1,85	2,01	-	2,48	-	-
5	PMAG	1,28	1,37	1,51	1,65	2,00	2,27	-
6	PMAG	1,30	1,39	1,51	1,65	1,99	2,28	-

Notes: * – Chemical shifts are given in ppm, all signals are revealed as broaden singlets. The first line revealing in strong field is the most intensive for signals of methyl and methylene groups.

** – There is the number of line for PAA and PAG in brackets.

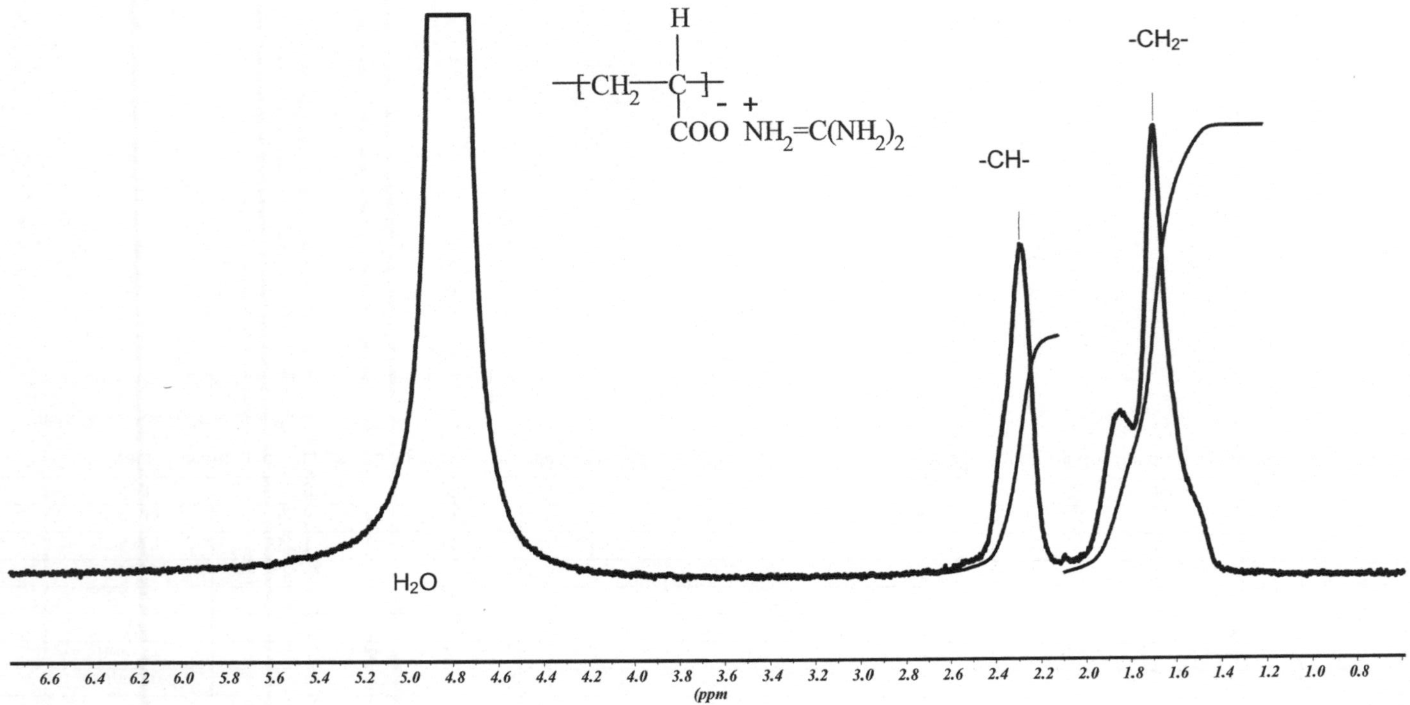

Figure 4.19. ^{1}H NMR spectrum of polyacrylate guanidine in D_2O.

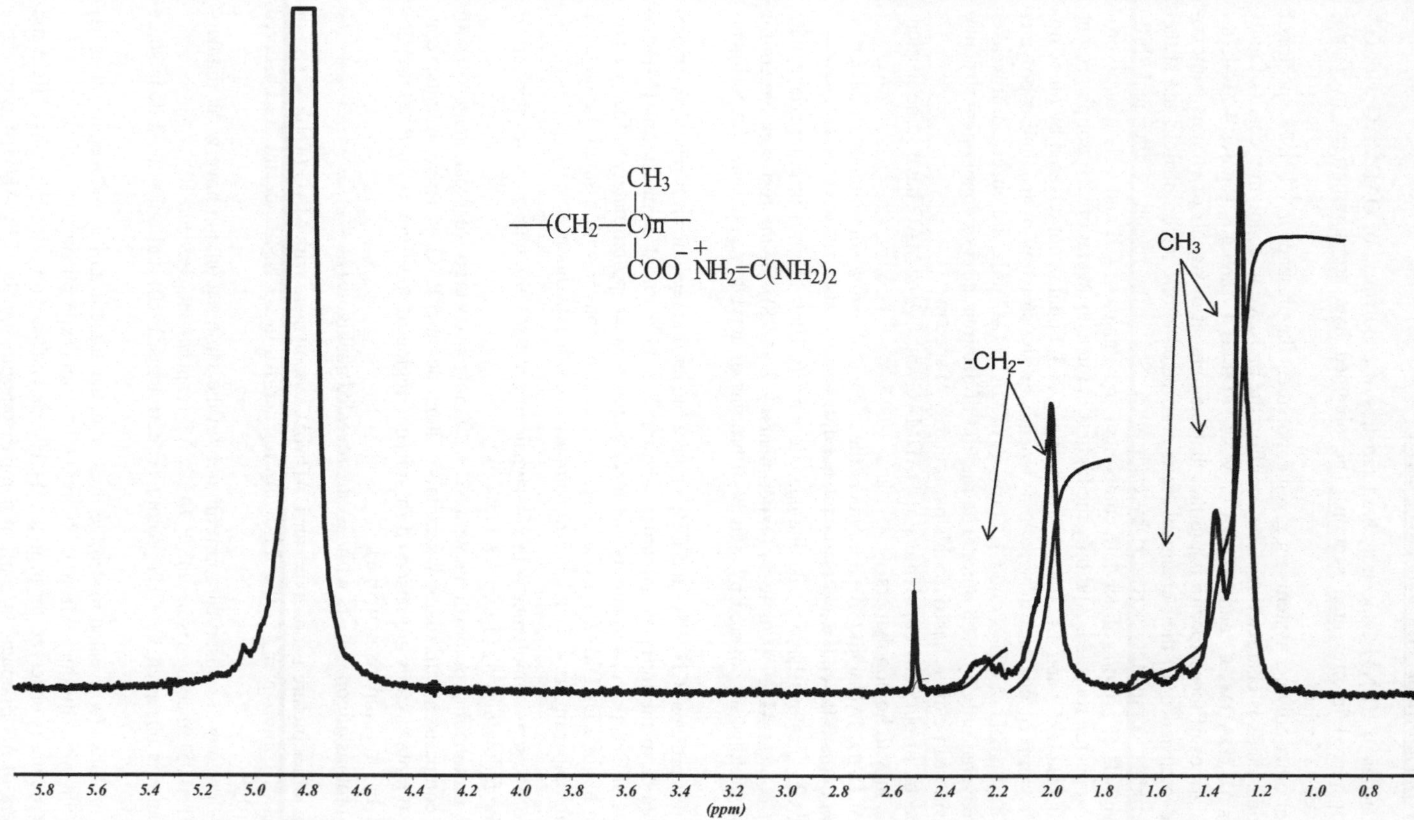

Figure 4.20. ^{1}H NMR spectrum of polymethacrylate guanidine in D_2O.

4.2.4.2. Copolymers with diallyldimethylammonium chloride

The particularities of ^{1}H NMR spectral characteristics of copolymers of DADMAC with GA and MAG are considered in presented section. The obtained data are summarized in Tables 4.7–4.14 and Figures 4.21–4.28.

As well as for other polymer systems for given copolymers the presence in ^{1}H NMR spectra of a number of broaden bands is characteristic; they confirm the presence in copolymer chain of units if both copolymers **M_1** (DADMAC) and **M_2** (GA or MAG) (Tables 4.7 and 4.8, Figures 4.21–4.28). One of the particularities of these spectra is the fact that the number of signals of various comonomer structures due to closeness of their chemical shifts is revealed in one region of field and at that they may be covered or completely screened by each other. So, in the region of chemical shifts 1,0–3,0 ppm the signals of protons of acrylate component and signals of CH and CH_2 groups' protons of DADMAC units are revealed. And only the highest in intensity signal of CH group's protons of DADMAC units (Table 4.7 and 4.8, lines 1–6) observing at 3,0 ppm is not covered by other signals. On the whole, form of this part of spectrum depends on the structure of resulted copolymer (Figures 4.21–4.28). Signals of CH_2N and CH_3 protons of DADMAC units are observed in weaker field and may be covered only by each other. The method of determination of copolymers structure which will be considered below is based on this particularity of spectra.

^{1}H NMR spectra of obtained copolymers of DADMAC and GA or MAG have some particularities in comparison with homopolymers.

In the case of DADMAC with GA copolymers the signals of diallyl comonomer are observed in ^{1}H NMR spectra in the following way (Table 4.7 and 4.9).

- The signals of CH_2 protons have weak diamagnetic shift (0,10–0,15ppm) in comparison with PDADMAC at content **M_1** ≥ 50 mole % (experiments 4–7 and 8); for the rest copolymers the signals of these protons are covered by signals of protons of acrylate comonomer (experiments 1–3).
- Signals of the rest protons (CH, CH_3 and CH_2N) have bigger diamagnetic shift (about 0,20ppm) at **M_1** ≥ 40 mole % in the initial comonomer mixture. For other initial compositions (**M_1** < 20 mole %) signals of CH_3 protons in copolymer spectra are revealed approximately in the same region of spectra as for PDADMAC, and of CH and CH_2N protons – with a small diamagnetic shift at 0,10ppm (experiments 4–7 and 1–3 in comparison with experiment 8).

In the case of the same copolymer with GA signals of acrylate comonomer are revealed in ^{1}H NMR spectra in the following way (Table 4.7 and 4.9)

- Signals of CH_2 protons for compositions with **M_1** ≤ 20 mole % unsergo weak paramagnetic shift (0,11–0,15) in comparison with analogous signals of homopolymer PAG. Whereas at higher content of **M_1** (≥ 40 mole %) they are observed in the same region of spectrum as for PAG (experiments 1–3 and 4–7 in comparison with 8).
- Signals of CH protons of copolymers with GA are revealed analogously.

Somewhat another picture is observed for DADMAC copolymers with MAG (Table 4.8). For signals of protons of comonomer **M_1** the following regularities in ^{1}H NMR spectra are observed (Table 4.8 and 4.10).

- Signals of CH_2 protons are observed approximately in the same region and have weak diamagnetic shift (0,01–0,05) in comparison with PDADMAC (experiments 3–9 and 10).
- Analogous picture is observed for CH_3 protons (experiments 2–10) and only the shift is some what bigger (0,05–0,11).
- Signals of CH protons are revealed in stronger field and diamagnetic shift is reduced as far as **M_1** is increased in the initial mixture – from 0,20 up to 0,05 ppm (experiments 3–8 and 10).
- The analogous picture is observed for protons of CH_2N: signals are revealed in stronger field and at higher content of **M_1** diamagnetic shift in comparison with signals of PDADMAC protons is decreased from 0,33 down to 0,06ppm (experiments 1–9).

Table 4.7. ^{1}H NMR spectral characteristics of copolymers of DADMAC (**M$_1$**) and GA (**M$_2$**)*.

No.	Comonomer	M$_1$							
	Group	-CH$_2$-		-CH-		CH$_3$-		-CH$_2$N-	
	M$_1$:M$_2$	1-1+1-2	1-3+1-4	1-5	1-6	1-7	1-8	1-9**	1-10
1	10:90	Closed by "**M$_2$**"		-	2,87	3,40	3,50	in 1-8	4,02
2	20:80	-*-		-	2,92	3,41	3,51	-*-	4,05
3	40:60	-*-		-	2,79	3,29	3,39	-*-	3,93
4	50:50	1,52	1,70	-	2,79	3,28	3,37	-*-	3,91
5	60:40	1,22+1,53	1,70	-	2,79	3,29	3,37	-*-	3,91
6	80:20	1,22+1,52	1,70	-	2,79	3,29	3,37	-*-	3,92
7	90:10	1,29+1,54	1,73	2,48	2,86	3,30	3,43	-*-	4,00
8	100:0	*1,41*+1,64	1,85+*1,98*	2,54	3,00	3,45	3,55	-*-	4,13

No.	Comonomer	M$_2$			M$_3$			
	Group	-CH$_2$-		-CH-	CH$_3$-	-CH$_2$N-	CH$_2$=	CH=
	M$_1$:M$_2$	2-1	2-2	2-3	3-1	3-2	3-3	3-4
9	0:100	1,69	1,82	2,28	(3,27)***	(4,14)***	(5,96)***	(6,30)***
1	10:90	1,80	1,96	2,39	3,30	4,17	6,01	6,42
2	20:80	1,82	1,97	2,43	3,30	4,17	6,01	6,32
3	40:60	1,69	1,85	2,27	3,18	4,06	5,90	6,20
4	50:50	1,70	1,84	2,28	3,16	4,03	5,88	6,19
5	60:40	1,70	1,85	2,29	3,16	4,04	5,88	6,19
6	80:20	1,70	1,86	2,29	3,17	4,03	5,88	6,18
7	90:10	1,73	1,89	2,31	3,21	"in 1-10"	5,92	6,22

Notes: * – All spectra were obtained in D_2O, chemical shifts are presented in ppm; in the second column the initial comonomer structure is presented;

** – The given line was covered by signal 1-8 or was revealed on this line in the shape of arm;

*** – The signals of monomer DADMAC.

Table 4.8. ^{1}H NMR spectral characteristics of copolymers of DADMAC (**M$_1$**) and MAG (**M$_2$**)*.

No.	Comonomer	M$_1$							
	Group	-CH$_2$-		-CH-**		CH$_3$-		-CH$_2$N-	
	M$_1$:M$_2$	1-1+1-2	1-3+1-4	1-5	1-6	1-7	1-8	1-9***	1-10
1	10:90	Covered by "**M$_2$**"		-	-	3,23	3,30	in 1-8	3,80
2	20:80	Covered by "**M$_2$**"		-	-	3,36	3,44	-*-	3,92
3	30:70	1,60	1,80	-	2,80 weak	3,38	3,44	-*-	3,92
4	40:60	1,61	1,84	-	2,82 double	3,40	3,47	-*- + 3,64	3,94
5	50:50	1,60	1,83	-	2,81 double	in 1-8	3,45	-*- + 3,62	3,94
6	60:40	1,62	1,84	-	2,85 double	3,40	3,47	-*- + 3,62	3,94
7	70:30	1,42+1,60	1,80	-	2,87 double	3,38	3,46	-*- + 3,82	4,01
8	80:20	1,44+1,60	1,81	-	2,88	in 1-8	3,47	-*- + 3,84	4,03
9	90:10	1,44+1,61	1,82	-	2,95	3,41	3,50	-*- 3,83	4,07
10	100:0	*1,41*+1,64	1,85+*1,98*	2,54	3,00	3,45	3,55	-*-	4,13

No.	Comonomer	M$_2$					M$_3$			
	Group	CH$_3$-			-CH$_2$-		CH$_3$-	-CH$_2$N-	CH$_2$=	CH=
	M$_1$:M$_2$	2-1	2-2	2-3+2-4	2-5	2-6	3-1	3-2	3-3	3-4
11	0:100	1,28	1,37	1,51+1,65	2,00	2,27	(3,27)****	(4,14) ****	(5,96)****	(6.30)****
1	10:90	1,12	1,21 weak	1,43	1,83	2,11	3,16	4,02	5,87	6.19
2	20:80	1,22	1,32 weak	1,52	1,95	2,16	3,25	4,13	5,97	6.27
3	30:70	1,25	1,37 weak	-	1,96	-	3,29	4,16	6,00	6.31
4	40:60	1,28	1,35 weak	-	2,03	-	3,31	4,17	6,03	6.33
5	50:50	1,27	Covered by "**M$_1$**"		2,04	-	3,29	4,16	6,00	6,31
6	60:40	1,30	-*-		2,05	2,16	3,29	4,17	6,01	6,31
7	70:30	1,27	-*-		2,00	2,15	3,27	4,15	5,98	6,30
8	80:20	1,30	-*-		2,00	2,15	Covered by "**M$_1$**"		5,99	6,32
9	90:10	1,31	-*-		2,00	2,18	3,28	-*-	5,99	6,32

Notes: * – All spectra were obtained in D_2O, chemical shifts are presented in ppm; in the second column the initial comonomer structure is presented;

** – Signals of CH protons were revealed as broaden weak singlets pr double lines or not revealed at all (-);

*** – The given line was covered by signal 1-8 or was revealed on this line in the shape of arm. For some structures the additional line was revealed for given group of signals;

**** – The signals of monomer DADMAC.

Table 4.9. The change of chemical shifts in copolymers DADMAC and GA in comparison with homopolymers*

No.	Comonomer	M₁					M₂	
	Group	$-CH_2-$		-CH-	CH_3-	$-CH_2N-$	$-CH_2-$	-CH-
	$M_1:M_2$	1-2	1-3	1-6	1-8	1-10	2-1	2-5
1	10:90	-	-	- 0,13	- 0,05	- 0,11	0,11	0,11
2	20:80	-	-	- 0,08	- 0,04	- 0,08	0,13	0,15
3	40:60	-	-	- 0,21	- 0,16	- 0,20	0,00	- 0,01
4	50:50	- 0,12	- 0,15	- 0,21	- 0,18	- 0,22	0,01	0,00
5	60:40	- 0,11	- 0,15	- 0,21	- 0,18	- 0,22	0,01	0,01
6	80:20	- 0,12	- 0,15	- 0,21	- 0,18	- 0,21	0,01	0,01
7	90:10	- 0,10	- 0,12	- 0,14	- 0,12	- 0,13	0,04	0,03

Note: * – For determination of chemical shift of corresponding signal from the value of copolymer we subtract the value for corresponding homopolymer; the negative value means diamagnetic shift of signal, and positive – paramagnetic shift of signal; ratio $M_1:M_2$ is presented for initial comonomer structure. The values "1-2" etc. indicate on corresponding line of spectrum.

Table 4.10. The change of chemical shifts in copolymers DADMAC and MAG in comparison with homopolymers*

No.	Comonomer	M_1					M_2	
	Group	$-CH_2-$		-CH-	CH_3-	$-CH_2N-$	CH_3-	$-CH_2-$
	$M_1:M_2$	1-2	1-3	1-6	1-8	1-10	2-1	2-5
1	10:90	-	-	-	- 0,25	- 0,33	- 0,16	- 0,17
2	20:80	-	-	-	- 0,11	- 0,21	- 0,06	- 0,05
3	30:70	- 0,04	- 0,05	- 0,20	- 0,11	- 0,21	- 0,03	- 0,04
4	40:60	- 0,03	- 0,01	- 0,18	- 0,08	- 0,19	0,00	0,03
5	50:50	- 0,04	- 0,02	- 0,19	- 0,10	- 0,19	- 0,01	0,04
6	60:40	- 0,02	- 0,01	- 0,15	- 0,08	- 0,19	0,02	0,05
7	70:30	- 0,04	- 0,05	- 0,13	- 0,09	- 0,12	- 0,01	0,00
8	80:20	- 0,04	- 0,04	- 0,12	- 0,08	- 0,10	0,02	0,00
9	90:10	- 0,03	- 0,03	- 0,05	- 0,05	- 0,06	0,03	0,00

Note: * – For determination of chemical shift of corresponding signal from the value of copolymer we subtract the value for corresponding homopolymer; the negative value means diamagnetic shift of signal, and positive – paramagnetic shift of signal; ratio $M_1:M_2$ is presented for initial comonomer structure. The values "1-2" etc. indicate on corresponding line of spectrum.

Table 4.11. Copolymers of DADMAC with GA: structure, calculation of composition and elemental analysis*.

No.	M_1:M_2 (initial)	M_1	M_2	M_3	M_4	M_5	C, %	H, %	N, %	Cl, %
1	0:100	-	100	-	-	-	36,64	6,92	32,04	-
		Elemental analysis results					36,60	6.90	29,85	-
2	10:90	10	85	5	-	-	40,43	7,36	28,49	2,06
		Elemental analysis results					40,85	7,95	26,35	1,95
3	20:80	18	72	10	-	-	43,83	7,76	25,29	3,97
		Elemental analysis results					43,45	8,10	23,15	3,60
4	40:60	34	52	14	-	-	48,63	8,35	20,60	7,48
		16	34	14	18	18	54,19	8,38	15,54	3,51
		Elemental analysis results					53,08	9,00	14,80	3,50
5	50:50	36	49	15	-	-	49,27	8,43	19,98	7,87
		18	31	15	18	18	54,93	8,47	14,55	3,95
		Elemental analysis results					54,20	9,60	13,00	3,95
6	60:40	39	49	12	-	-	49,22	8,44	19,90	8,49
		19	29	12	20	20	55,31	8,48	14,04	4,14
		Elemental analysis results					55,65	9,40	13,00	4,15
7	80:20	55	38	7	-	-	15,38	8,81	17,36	12,08
		38	21	7	17	17	56,53	8,82	12,40	8,33
		Elemental analysis results					56,05	10,00	10,90	8,30
8	90:10	75	22	3	-	-	54,70	9,30	13,69	16,35
		63	10	3	12	12	58,33	9,31	10,20	13,82
		Elemental analysis results					56,60	9,90	9,11	13,82

Notes: * – In experiment 1: the first line presents calculation by formula, the second – elemental analysis results; in experiments 2-3: the first line presents the calculation by scheme "M_1–M_3", the second – elemental analysis results; in experiments 4-8: the first line presents the calculation by scheme "M_1–M_3", the second – the calculation by scheme "M_1–M_5", the third – elemental analysis results.

Table 4.12. Copolymers of DADMAC with MAG: structure, calculation of composition and elemental analysis*.

No.	M_1:M_2 (initial)	M_1	M_2	M_3	M_4	M_5	C, %	H, %	N, %	Cl, %
1	0:100	-	100	-	-	-	41,37	7,64	28,95	-
		Elemental analysis results					40,95	8,00	27,60	-
2	10:90	3	94	3	-	-	42,74	7,78	27,68	0,53
		Elemental analysis results					41,45	8,15	26,65	0,50
3	20:80	5	88	7	-	-	44,15	7,92	26,35	1,16
		Elemental analysis results					44,00	8,30	24,45	1,20
4	30:70	11	78	11	-	-	46,30	8,15	24,30	2,83
		Elemental analysis results					46,95	8,30	22,30	3,10
5	40:60	24	65	11	-	-	48,54	8,45	21,73	4,52
		8	49	11	16	16	53,62	8,62	17,49	1,71
		Elemental analysis results					52,80	9,10	16,55	1,70
6	50:50	28	55	17	-	-	51,07	8,71	19,41	6,14
		12	39	17	16	16	56,06	8,88	15,24	2,70
		Elemental analysis results					55,90	9,35	14,40	2,70
7	60:40	34	55	11	-	-	50,36	8,64	19,68	7,43
		15	36	11	19	19	56,50	8,90	14,56	3,29
		Elemental analysis results					55,75	9,70	13,65	3,30
8	70:30	43	51	6	-	-	50,63	8,77	19,01	9,32
		17	25	6	26	26	58,99	9,06	12,04	3,55
		Elemental analysis results					58,60	10,00	11,00	3,55
9	80:20	50	48	2	-	-	50,97	8,86	18,33	10,94
		22	20	2	28	28	59,84	9,16	10,93	4,82
		Elemental analysis results					58,50	9,90	10,15	4,82
10	90:10	61	36	3	-	-	53,30	9,15	15,80	13,44
		46	21	3	15	15	58,13	9,21	11,80	10,11
		Elemental analysis results					57,85	9,80	10,78	10,10

Notes: * – In experiment 1: the first line presents calculation by formula, the second – elemental analysis results; in experiments 2-4: the first line presents the calculation by scheme "M_1–M_3", the second – elemental analysis results; in experiments 5-10: the first line presents the calculation by scheme "M_1–M_3", the second – the calculation by scheme "M_1–M_5", the third – elemental analysis results.

Table 4.13. Copolymers of DADMAC with MAG: structure, calculation of composition and elemental analysis (high conversion)*.

No.	M_1:M_2 (initial)	M_1:	M_2	M_3	M_4	M_5	C, %	H, %	N, %	Cl, %
1	30:70 (dialysis)	26	69	5	-	-	47,35	8,36	22,62	5,64
		9	52	5	17	17	52,75	8,54	18,11	1,91
		Elemental analysis results					52.25	9.30	16,75	1,90
2	30:70 (precipitation into acetone)	26	64	10	-	-	48,86	8,50	21,34	5,77
		24	67	9*	-	-	47,26	8,40	22,33	7,15
		Elemental analysis results					47.20	9,30	22,14	8,15
3	70:30 (dialysis)	56	39	5	-	-	52,85	9,07	16,43	12,35
		27	10	5	29	29	62,04	9,38	8,77	6,01
		Elemental analysis results					60.80	11,10	6,70	6,00
4	70:30 (precipitation into acetone)	63	26	11	-	-	55,69	9,38	13,70	13,90
		57	33	10*	-	-	53,45	9,20	15,37	14,67
		Elemental analysis results					53.80	10,40	14,30	14,00

Notes: * – In experiments 1 and 3: the first line presents the calculation by scheme "M_1–M_3", the second – the calculation by scheme "M_1–M_5", the third – elemental analysis results; in experiments 2 and 4: the first line presents the calculation by scheme "M_1–M_3", the second – the calculation where DADMAC was considered as impurity (*), the third – elemental analysis results.

Table 4.14. The composition of copolymers of DADMAC with GA and MAG by schemes "M_1–M_3" and "M_1–M_5".

No.	M_1:M_2 *	GA or MAG	Copolymer structure									
			For r_1 and r_2**		Scheme "M_1 – M_3"**			Scheme "M_1 – M_5"***				
			M_1	M_2	M_1	M_2	M_3	M_1	M_2	M_3	M_4	M_5
1	10:90	GA	10	90	10	85	5	10	85	5	-	-
		MAG	3	97	3	94	3	3	94	3	-	-
2	20:80	GA	18	82	18	72	10	18	72	10	-	-
		MAG	5	95	5	88	7	5	88	7	-	-
3	30:70	GA	-	-	-	-	-	-	-	-	-	-
		MAG	11	89	11	78	11	11	78	11	-	-
4	40:60	GA	34	66	34	52	14	16	34	14	18	18
		MAG	24	76	24	65	11	8	49	11	16	16
5	50:50	GA	36	64	36	49	15	18	31	15	18	18
		MAG	28	72	38	55	17	12	39	17	16	16
6	60:40	GA	39	61	39	49	12	19	29	12	20	20
		MAG	34	66	34	55	11	15	36	11	19	19
7	70:30	GA	-	-	-	-	-	-	-	-	-	-
		MAG	43	57	43	51	6	17	25	6	26	26
8	80:20	GA	55	45	55	38	7	38	21	7	17	17
		MAG	50	50	50	48	2	22	20	2	28	28
9	90:10	GA	75	25	75	22	3	63	10	3	12	12
		MAG	61	39	61	36	3	46	21	3	15	15

Notes: * – The structure of initial reaction mixture;
** – By the data of ^{1}H NMR spectroscopy;
*** – By the data of ^{1}H NMR spectroscopy and elemental analysis.

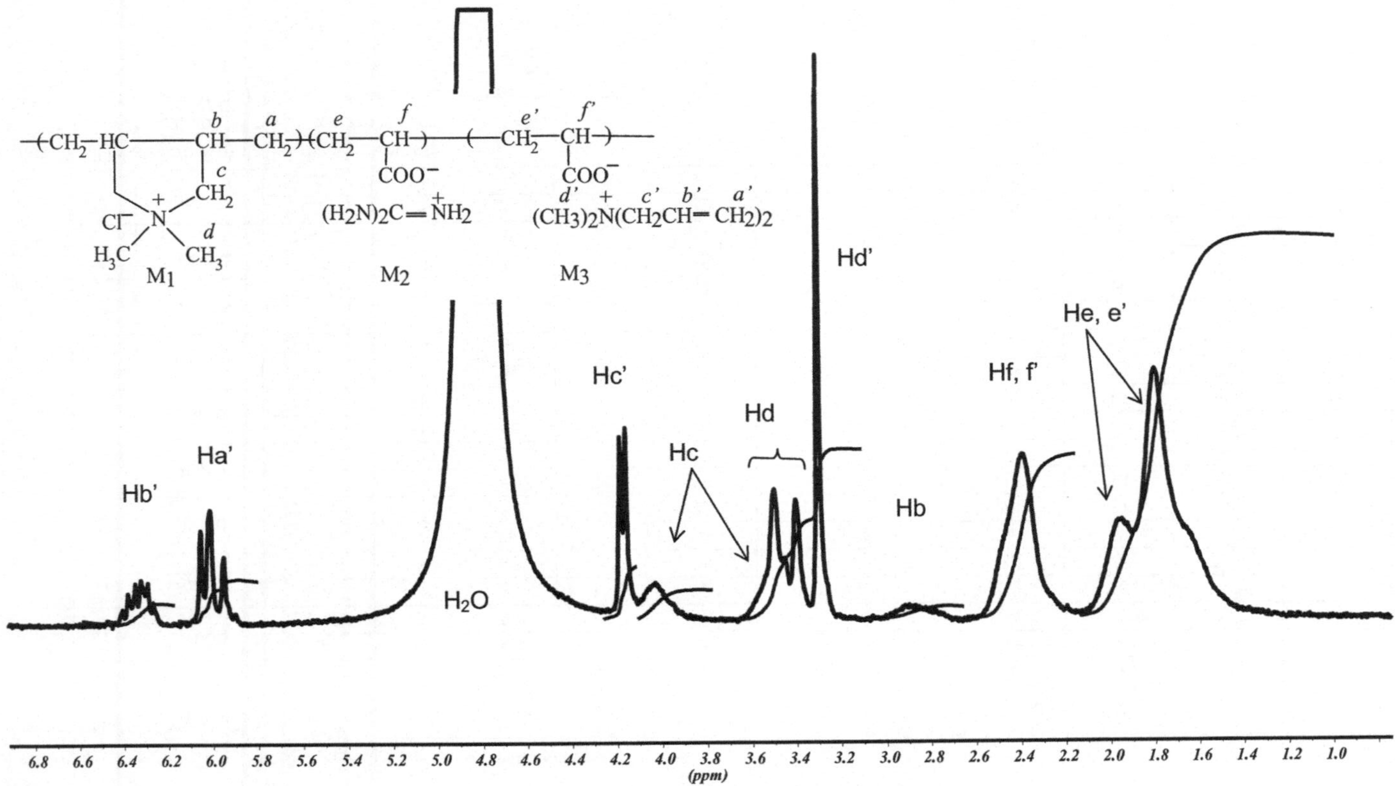

Figure 4.21. ^{1}H NMR spectrum of copolymer DADMAC and GA (initial monomer composition was 10 : 90) in D_2O.

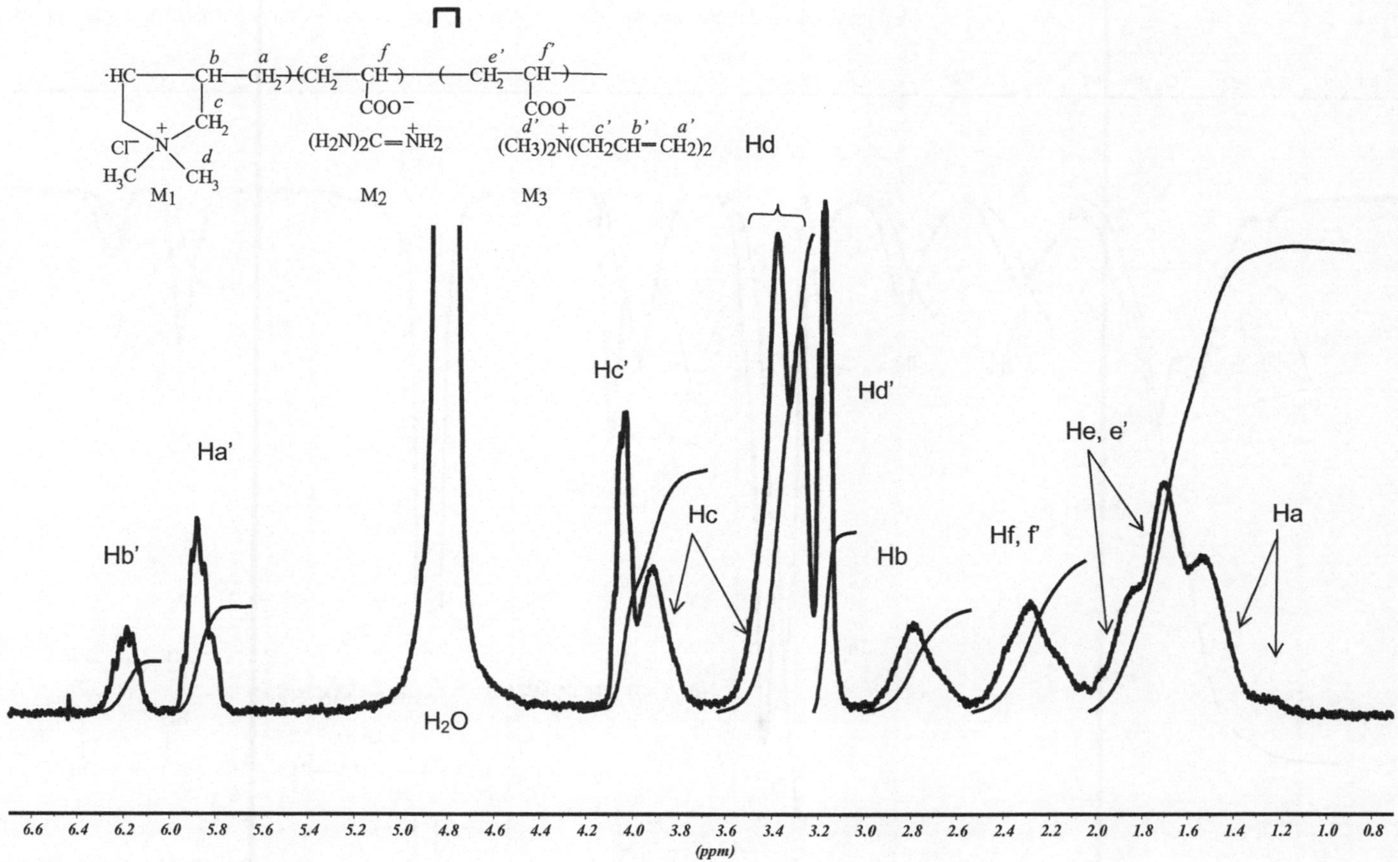

Figure 4.22. ^{1}H NMR spectrum of copolymer DADMAC and GA (initial monomer composition was 50 : 50) in D_2O.

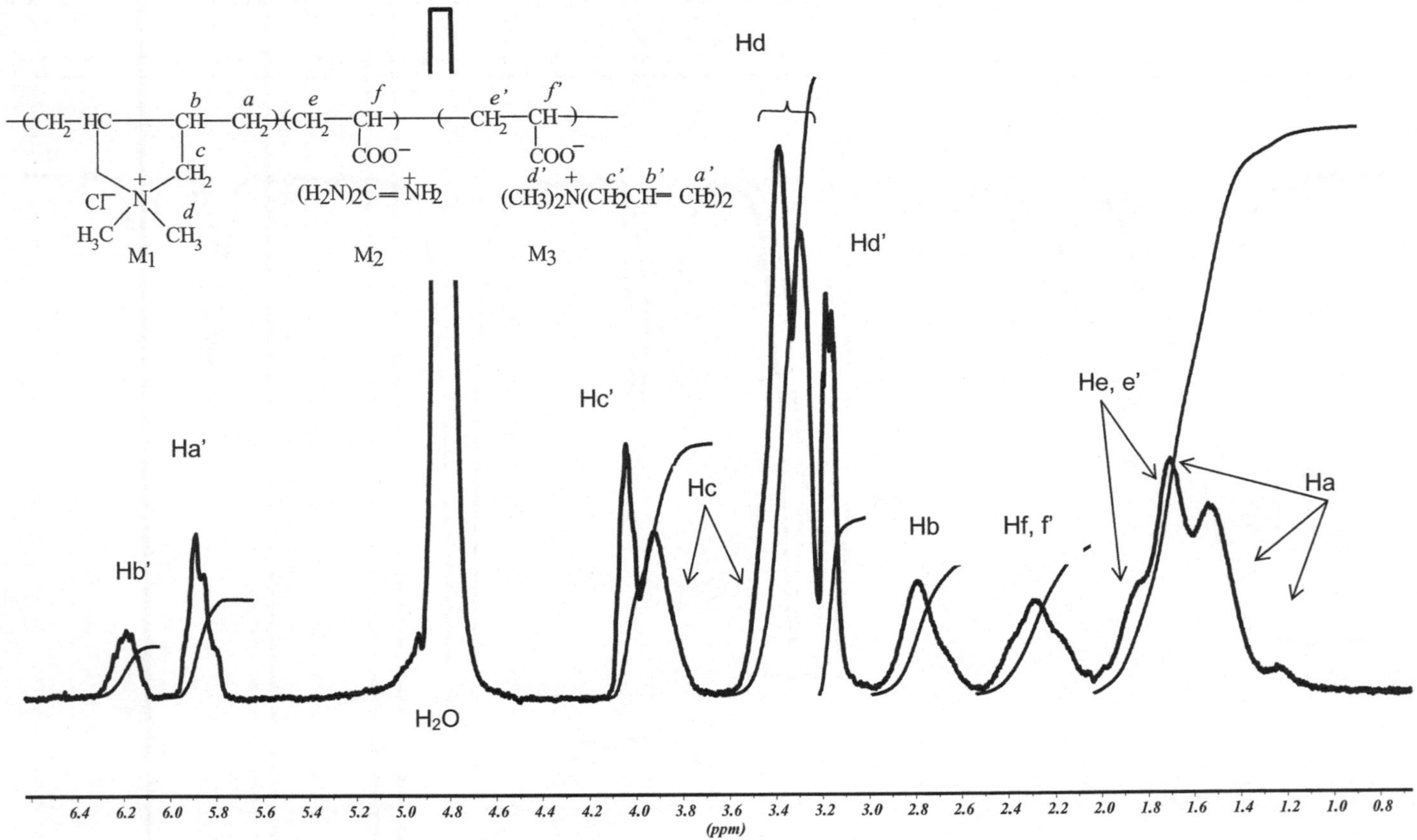

Figure 4.23. ^{1}H NMR spectrum of copolymer DADMAC and GA (initial monomer composition was 60 : 40) in D_2O.

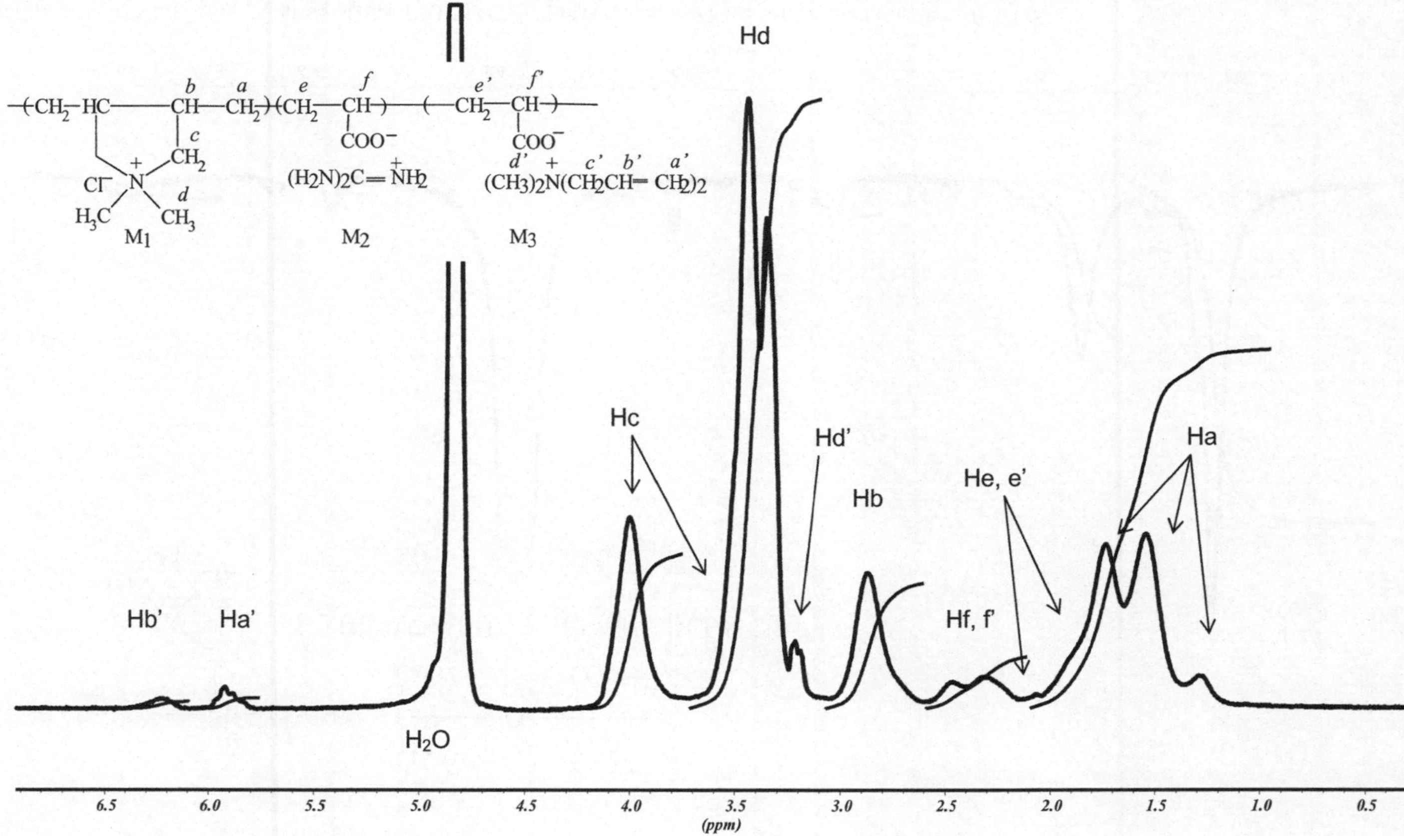

Figure 4.24. ^{1}H NMR spectrum of copolymer DADMAC and GA (initial monomer composition was 90 : 10) in D_2O.

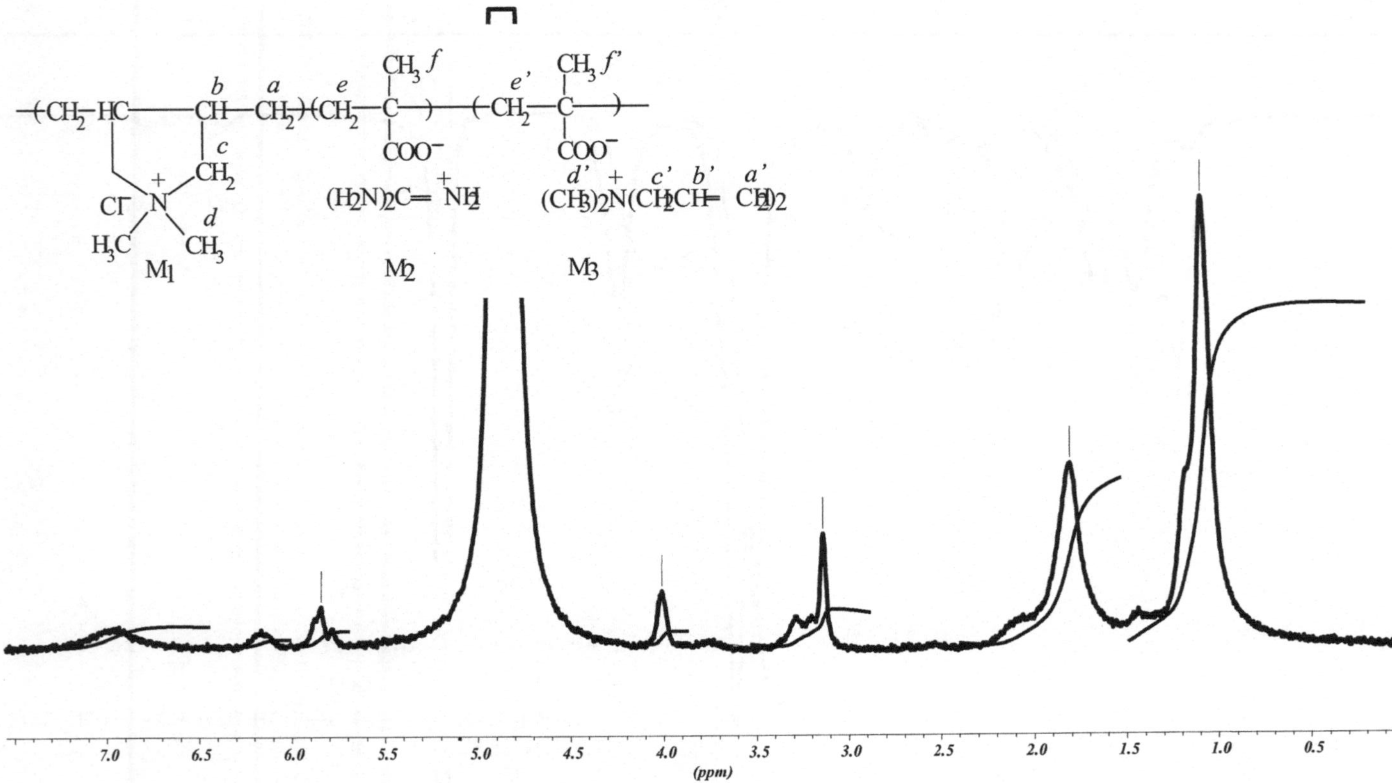

Figure 4.25. ^{1}H NMR spectrum of copolymer DADMAC and MAG (initial monomer composition was 10 : 90) in D_2O.

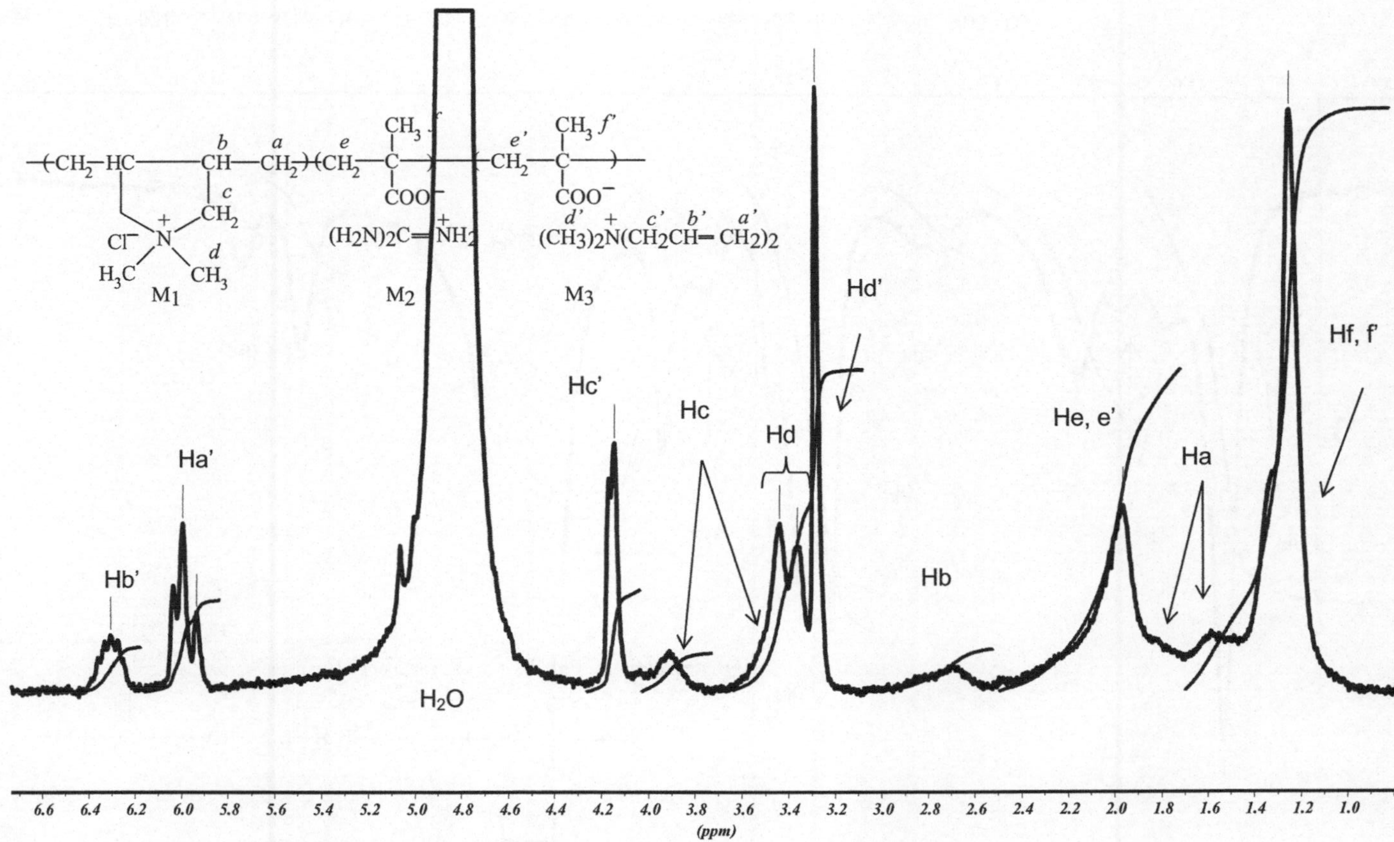

Figure 4.26. ^{1}H NMR spectrum of copolymer DADMAC and MAG (initial monomer composition was 30 : 70) in D_2O.

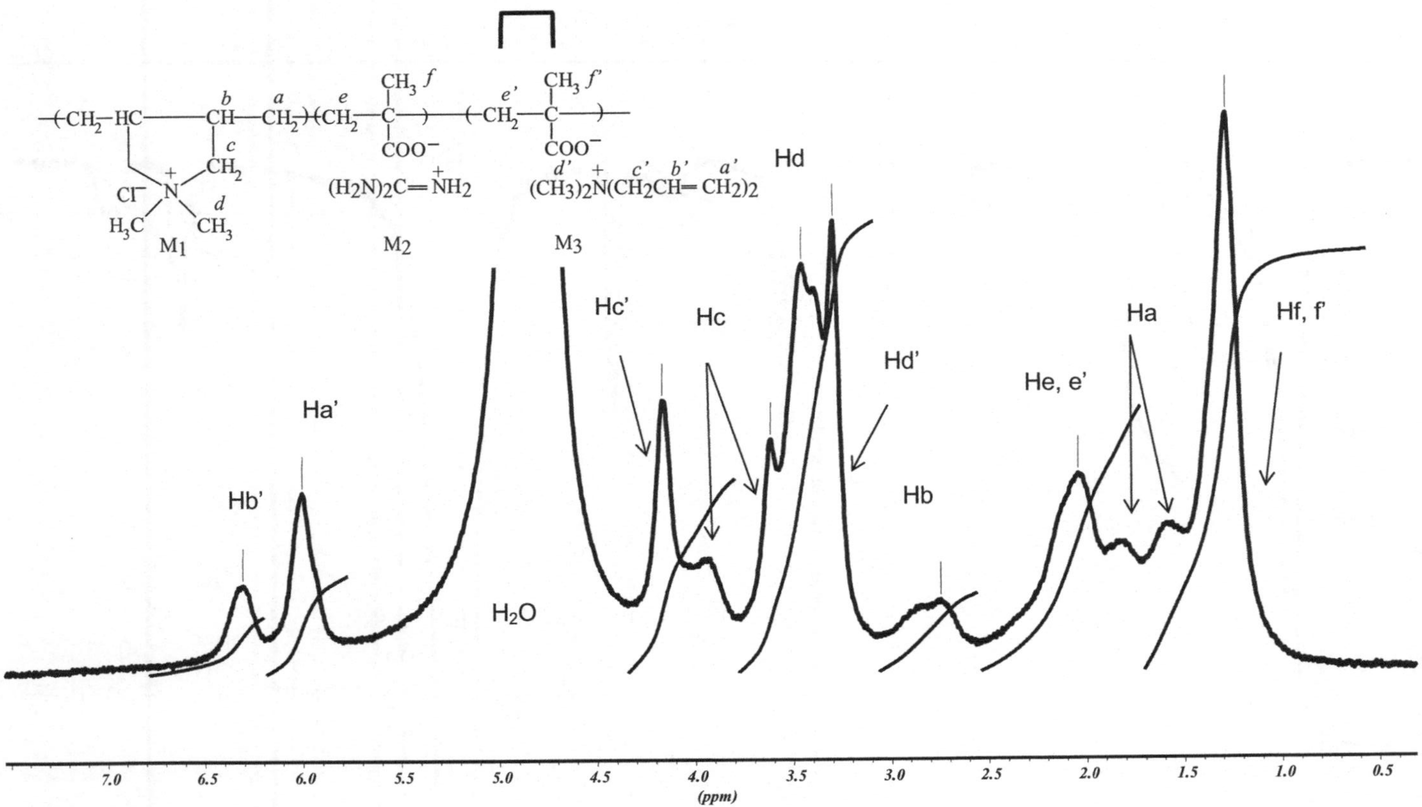

Figure 4.27. ^{1}H NMR spectrum of copolymer DADMAC and MAG (initial monomer composition was 50 : 50) in D_2O.

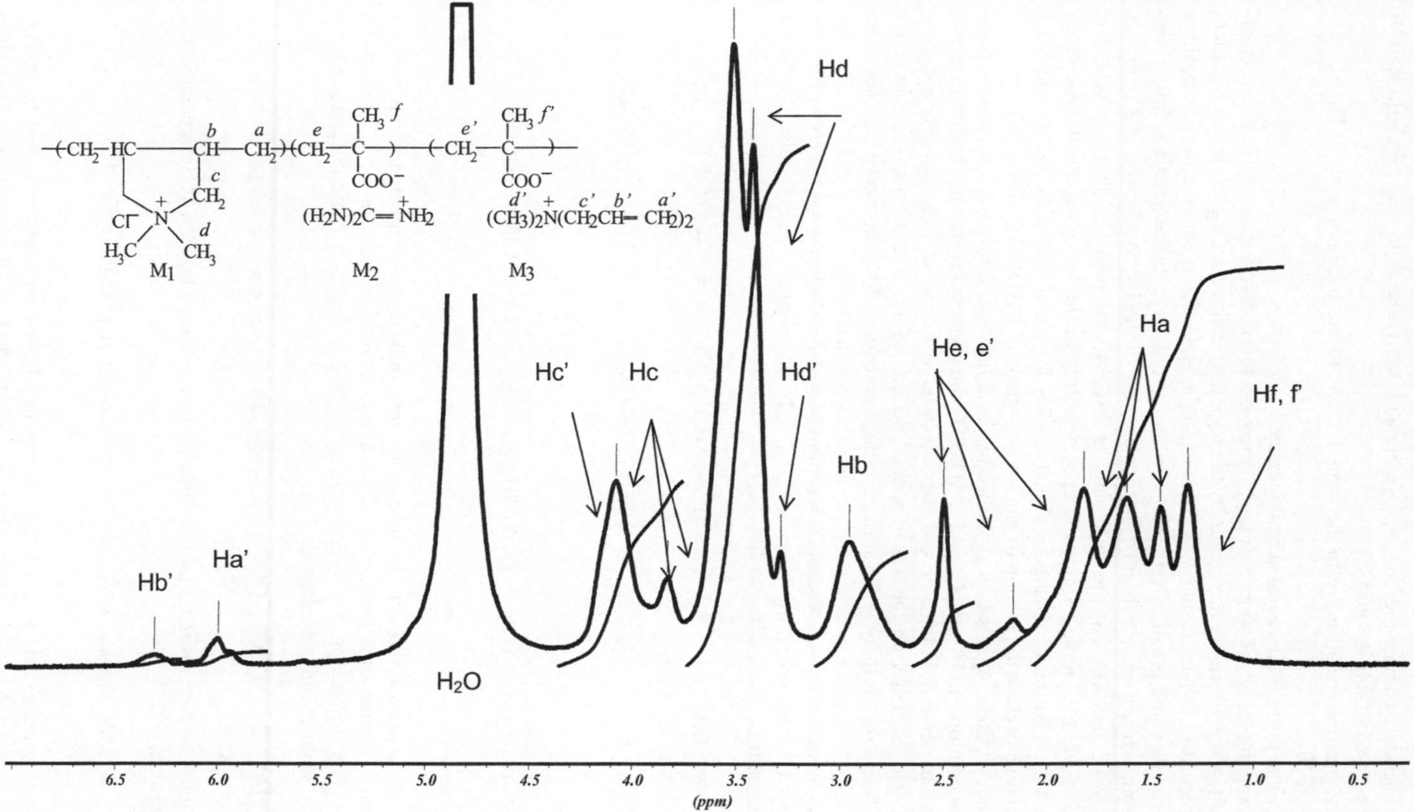

Figure 4.28. ^{1}H NMR spectrum of copolymer DADMAC and MAG (initial monomer composition was 90 : 10) in D_2O.

For signals of acrylate comonomer for copolymers with MAG the following regularities in ^{1}H NMR spectra are observed (Table 4.8 and 4.10).

- Signals of methyl protons are observed in the same field as for analogous signals in PMAG (experiments 2–9 and 10).
- Signals of methylene protons are revealed analogously.

While analyzing the ^{1}H NMR spectra of copolymers we found a very characteristic particularity: the presence of signals which are analogous by the region of their revealing to signals of monomer DADMAC. Intensity of these signals was changed in dependence on comonomer composition of initial reaction mixture and practically in all cases they didn't have characteristic splitting connected with spin-spin interaction. Position of protons' signals of diallyl counter-ion in $\mathbf{M_3}$ for copolymers of DADMAC and GA are the following: at $\mathbf{M_1} \geq 40$ mole % signals are revealed in stronger field, and at $\mathbf{M_1} \geq 20$ mole % they are approximately in the same region of field as for DADMAC (Table 4.4). Position of analogous signals in $\mathbf{M_3}$ for copolymers DADMAC and MAF for all groups of protons is close to position of analogous signals of DADMAC (Table 4.8)

The appearance of these signals was explained by the following. While studying kinetic regularities of copolymerization of GA (MAG) and DADMAC the conversion was not higher than 5%. After this moment separation of obtained copolymers from large amount of initial comonomers was carried out by dialysis method with the use of special packets. Since both comonomers have ion structure, so in dialysis process there is the shift of various reactions of ion exchange between ionogenic compounds which in water solution present as solvate separated ion pairs and ions (Scheme 4.3). In this Scheme in two following exchange reactions and also in Scheme 4.4 which will be presented below **M** means corresponding monomer, and "**M**" in inverted commas relates to corresponding copolymer unit.

1) $\underset{M1}{DA^{+}Cl^{-}} \rightleftharpoons DA^{+} + Cl^{-}$

3) $\underset{\text{"M1"}}{\sim\sim\overset{+}{N}(CH_3)_2\,Cl^{-}} \rightleftharpoons \sim\sim\overset{+}{N}(CH_3)_2 + Cl^{-}$

2) $\underset{M2}{G^{+}\,MA^{-}} \rightleftharpoons G^{+} + MA^{-}$

4) $\underset{\text{"M2"}}{G^{+}\,{}^{-}OOC\sim\sim} \rightleftharpoons G^{+} + {}^{-}OOC\sim\sim$

$DA^{+} = (CH_2{=}CHCH_2)_2N^{+}(CH_3)_2 \qquad G^{+} = (NH_2)_2C{=}N^{+}H_2 \qquad MA^{-} = CH_2{=}C(R)COO^{-}$

Scheme 4.3.

Low-molecular ions pass through the membrane into outside water medium from both comonomers and chain counter-ions. Formation of guanidine hydrochloride as a result of exchange leads to formation of modified units $\mathbf{M_3}$ by appearance of which in polymer chain the appearance of signals of additional counter-ion is explained:

$$\underset{M1}{DA^{+}\,Cl^{-}} + \underset{\text{"M2"}}{G^{+}\,{}^{-}OOC\sim\sim} \rightleftharpoons \underset{\text{"M3"}}{DA^{+}\,{}^{-}OOC\sim\sim} + G^{+}Cl^{-}$$

We should note that there were no units of another type when on DADMAC unit there was an acrylate anion as counter-ion:

$$\underset{M2}{G^{+}\,MA^{-}} + \underset{\text{"M1"}}{\sim\sim\overset{+}{N}(CH_3)_2\,Cl^{-}} \rightleftharpoons \sim\sim\overset{+}{N}(CH_3)_2\,MA^{-} + G^{+}Cl^{-}$$

The analysis of spectral data showed that content of $\mathbf{M_3}$ in copolymer extracted by dialysis passed through the maximum which in both cases was observed at initial comonomer composition 50 : 50 (for GA – 15%, for MAG – 17%, Table 4.14, experiment 5).

To demonstrate the fact that formation of $\mathbf{M_3}$ occurs during dialysis process and that copolymerization constants may be calculated by the ratio $\mathbf{M_1} : \mathbf{M_2}$ (where $\mathbf{M_1}$ – the sum of acrylate units with various counter-ions) we carried out copolymerization of MAG and DADMAC with high conversion for two structures – 30:70 and 70:30. Reaction masses were divided in two parts when reaction stopped; and copolymer in one case was extracted by dialysis, and in another was precipitated into acetone. It turned out at that that in the case of precipitation the copolymers containing two comonomers with some impurity (about 10%) of initial DADMAC were formed, and more active comonomer MAG in accordance with ^{1}H NMR spectroscopy data was absent (Table 4.13, experiments 2 and 4).

Taking into account all said above the technique of calculation of comonomer structure of copolymers was developed. The following facts were considered while creating this technique (see Figures 4.21–4.28)

Integrals of signals of CH_2=($H_{a'}$) and CH= ($H_{b'}$) of protons of counter-ion of diallyldimethylammonium (DADMA) revealing separately in weak field near to 6,0ppm correspond to 6 protons and allow by equation 8 by integral intensities determining pf specific (molar) part $\mathbf{M_3}$:

$$"1H_{M3}" = ("H_{a'}": 4 + "H_{b'}": 2) : 2 \quad (8)$$

Integrals of signals revealing in the region 3,0–4,5ppm correspond to 4 protons of CH_2N ($H_{c'}$) and 6 protons of CH_3 ($H_{d'}$) of counter-ion groups of DADMA, and also 10H of diallyl unit to 4 protons of CH_2N (H_c) and 6 protons of CH_3 (H_d) groups. So, by equation 9 we may dxetermine mole part $\mathbf{M_1}$:

$$"1H_{M1}" = [(\Sigma\ "H_c" + "H_d" + "H_{c'}" + "H_{d'}") - ("1H_{M3}" \times 10)] : 10 \quad (9)$$

Integrals of signals revealing in strong field 1,0–3,0ppm correspond to 4 protons of CH_2 (H_a) and 2 protons of CH (H_b) of groups of diallyl comonomer $\mathbf{M_1}$ and to 3 protons of CH_2 ($H_{e,\,e'}$) and CH ($H_{f,\,f'}$) in the case of GA or to 5 protons of CH_2 ($H_{e,\,e'}$) and CH_3 ($H_{f,\,f'}$) in the case of MAG. On the base of this assignment by equation 10 we may determine mole part of acrylate comonomer $\mathbf{M_1}$:

$$"1H_{M2}"\ (GA) = [(\Sigma\ "H_{e,\,e'}" + "H_{f,\,f'}" + "H_a" + "H_b") - ("1H_{M1}" \times 6)] : 3 \quad (10a)$$

$$"1H_{M2}"\ (MA\Gamma) = [(\Sigma\ "H_{e,\,e'}" + "H_{f,\,f'}" + "H_a" + "H_b") - ("1H_{M1}" \times 6)] : 5 \quad (10b)$$

Basing on obtained values of molar parts (equations 8–10) we may calculate the content of comonomers in copolymer in molar percents (equations 11 and 12):

$$M_1\ (\text{моль. \%}) = "1H_{M1}" : ("1H_{M1}" + "1H_{M2}") \times 100\% \quad (11)$$

$$M_2\ (\text{моль. \%}) = "1H_{M2}" : ("1H_{M1}" + "1H_{M2}") \times 100\% \quad (12)$$

The data of these calculations are presented in Table 4.14. Content of $\mathbf{M_3}$ entering into $\mathbf{M_2}$ may be determined by the equation 13:

$$M_3\ (\text{mole \%}) = "1H_{M3}" : ("1H_{M1}" + "1H_{M2}") \times 100\% \quad (13)$$

Copolymerization constants for system DADMAC–GA were calculated for all compositions (Table 4.14, experiments 1, 2, 4–6, 8, 9, columns 4 and 5). Under copolymerization of system DADMAC–MAG at the content of acrylate component in initial reaction mixture ≥ 70% reaction solution was heterogeneous, so when calculating $\mathbf{r_1}$ and $\mathbf{r_2}$ for given pair of comonomers we used values obtained for content of MAG ≤ 60% in the initial reaction solution (Table 4.14, experiments 4–9, columns 4 and 5).

Analysis of ion exchange mentioned in the explanation of appearance in extracted polymers of $\mathbf{M_3}$ set us proposing that this process is more complex and formation of structural units in copolymer which are not revealed in ^{1}H NMR spectra is possible. In other words the exchange between ionogenic units of copolymer with following extraction of low-molecular counter-ions into outside water medium during the dialysis process may occur. At that in equal amount the $\mathbf{M_4}$ and $\mathbf{M_5}$ units are formed which become the pair of counter-ions (Scheme 4.4). Conservation of copolymers solubility at that indicates on the fact that such pairs of counter-ions are appeared preliminary intramolecularly. This conclusion was confirmed by additionally carried out experiment when water solutions of PDADMAC and PMAG (PAG) homopolymers were mixed and insoluble polyelectrolyte complex was obtained.

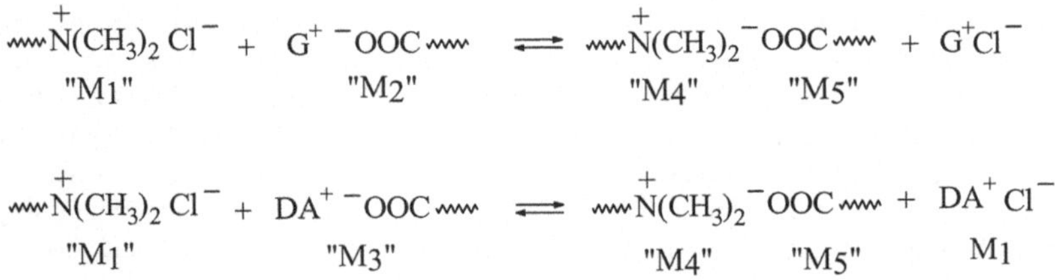

Scheme 4.4.

Combined analysis of data of ^{1}H NMR spectroscopy and elemental analysis confirmed our supposition. As it obvious from elemental analysis results (Table 4.11 and 4.12) for a number of comonomers structures the calculations by scheme "$\mathbf{M_1}$–$\mathbf{M_3}$" don't correspond to elemental composition obtained by elemental analysis – the carbon content in calculation is reduced, and nitrogen and chlorine are increased that is easily explained by scheme "$\mathbf{M_1}$–$\mathbf{M_5}$", since under extraction of $\mathbf{M_4}$ and $\mathbf{M_5}$ during dialysis the guanidinehydrochloride and DADMAC are transferred into outside water medium that leads to the decrease of nitrogen and chlorine content and to the increase of carbon part. Since chlorine content in copolymer is directly connected with amount of $\mathbf{M_1}$ then we may determine the content of $\mathbf{M_1}$ units (nov.) by chlorine content reduce:

$$M_1\ (\text{nov.}) = M_1\ (\text{NMR}) \times \text{Cl(el. an.)} : \text{Cl(NMR)},$$

and as M_1 (NMR) = M_1 (nov.) + M_4, si we may calculate the amount of $\mathbf{M_4}$ and equal to it $\mathbf{M_5}$, then M_2 (nov.) = M_2 (NMR) – M_4.

Further recalculation of elemental composition by scheme "$\mathbf{M_1}$–$\mathbf{M_5}$" gives good coincidence of results of calculation and elemental analysis results (Table 4.11 and 4.12).

We also showed (Table 4.13) that under copolymerization with deep conversion degrees for a half of reaction mixture which was dialyzed one should use scheme "$\mathbf{M_1}$–$\mathbf{M_5}$". For the second half of reaction mixture which was precipitated the best results were obtained by scheme "$\mathbf{M_1}$–$\mathbf{M_2}$" and revealing in ^{1}H NMR spectra DADMAC should be considered as impurity. Comparison of ^{1}H NMR data and elemental analysis results for outside water fractions of dialysis which were studied in given case indicated on significant part of GHC in their structure. It was expressed in small content of carbon and high content of nitrogen and chlorine in comparison with calculations carried out by NMR data, since in spectrum obtained in water the signals of protons of GHC were not revealed as a result of too fast exchange.

We should note, that amount of $\mathbf{M_4}$ and $\mathbf{M_5}$ determined by elemental analysis data (consideration of less amount of chlorine) is slightly deviates from real value since the content of various elements during elemental analysis is determined with error 3–5%, i.e. in such calculation real content of units $\mathbf{M_1}$ and $\mathbf{M_2}$ is somewhat reduced, and of $\mathbf{M_4}$ and $\mathbf{M_5}$ it is somewhat increased. As it is obvious from data of Table 4.14 the maximum content of units $\mathbf{M_4}$ and $\mathbf{M_5}$ corresponds to those copolymers in which the ratio $\mathbf{M_1} : \mathbf{M_2}$ is close to 50 : 50 (for GA – experiment 6, and for MAG – experiments 7 and 8).

One more important result obtained under combined analysis of ^{1}H NMR data and elemental analysis results is in the fact that at high content of acrylate component in the initial comonomer mixture $[\mathbf{M_2}] \geq 70$ mole % the composition of resulted copolymers with good coincidence is calculated by the scheme "$\mathbf{M_1}$–$\mathbf{M_3}$" (Table 4.11, experiments 2, 3; Table 4.12 experiments 2–4). In these cases t is obviously connected with formation of large segments of acryl comonomer which prefer to form spatial structures of ion and hydrogen interaction characteristic for homopolymer; the picture is only supplemented by formation of $\mathbf{M_3}$ during dialysis. The difference in the case of composition $[\mathbf{M_1}]:[\mathbf{M_2}]$ = 30:70 (for MAG) which was studied at deep conversion degrees and treated during dialysis was connected with the fact that in this case the gradient copolymer was formed at the expense of gradual decrease of content of more active MAG in copolymerization process. At that the resulted segments consisting of DADMAC units long enough which participate in ion exchange under dialysis with formation of $\mathbf{M_4}$ and $\mathbf{M_5}$ (Table 4.13, experiment 1 and Table 4.12, experiment 5; it is important to note that these two copolymers have close structure calculated by scheme "$\mathbf{M_1}$–$\mathbf{M_3}$").

Now we want to analyze the position of signals of copolymer obtained under investigation of kinetics of initial reaction solution of the composition 10 : 90 for pair DADMAC–MAG (Table 4.8, experiment 1) which is out of general regularities: in ^{1}H NMR spectrum of this copolymer signals of all protons groups are revealed in stronger field. It may be explained as follows.

For correct determination of copolymer structure with low content of diallyl comonomer under the change of proton spectrum we had to prepare more concentrated solution: instead of 5–7 mass % as was made for other copolymers, we prepared the solution with copolymer content 15 mass %. If concentration is increased the ball-shaped conformation of copolymer acrylate part becomes more compact that leads to decrease of dis-screening effect of carboxyl groups (for confirmation of this supposition we prepared 25% solution and signals of all protons were shifted to even stronger field, please, compare Figure 4.25 and Table 4.8 experiment 1). We also may propose that the insufficient amount of water molecules participating in hydrogen bonding penetrate into internal sphere and the broad band at 7ppm characteristic for guanidine counter-ion amine protons appearing due to incomplete exchange hydrogen–deuterium in this case indicates on it. Amine protons were not revealed in any other copolymers or homopolymers spectra (Figures 4.21–4.28).

The differences in chemical shifts for homo- and copolymers for corresponding groups of protons are not significant, but they have definite tendencies (Tables 4.9 and 4.10). They may be explained in the following way. Chemical shifts in copolymers' acrylic units practically maintain their values characteristic for homopolymers that confirms the analogous to homopolymers structure and environment of these units in copolymers. Differences for copolymers obtained at high initial content of acrylic comonomer (≥ 80 mole %) also confirm the tendency of structure remaining peculiar to homopolymer. Inclusion into copolymers structures of small amount of DADMAC units in the case of GA leads to formation of more flocculent ball-shaped structure that strengthens the dis-screening action of carboxyl groups and leads to paramagnetic shift of signals of acrylate comonomer protons (Table 4.7, experiments 1 and 2). In the case of MAG vise versa more compact structure is formed (that even leads to formation of solid phase in reaction mass) and reduction of dis-screening influence of carboxyl groups causes diamagnetic shift (Table 4.8, experiments 1 and 2). Due to the same reasons the signals of diallyl comonomer undergo diamagnetic shift: signals of CH_3 and CH_2N protons first of all. Increase of DADMAC units content in copolymer composition and also formation of polymer counter-ions first of all influence on positions of signals of diallyl comonomer which are shifted into stronger field in comparison with signals of PDADMAC at the expense of dis-screening action reduction of quaternizated nitrogen. Most of all this fact effects on signals of CH_3 and CH_2N protons and to a less degree on signals of CH and CH_2 protons (especially in the case of copolymers with MAG).

4.3. Thermo-physical characteristics of synthesized products

The thermo-stability was studied by methods of differential thermal analysis (DTA) and differential scanning calorimetry (DSC) in air and in argon. The behavior of synthesized monomers (DAGA, DAGTFA, GA, MAG), homopolymers (PAG, PMAG, PDADMAC was also studied for comparison) was studied. The number of copolymers of listed monomers and DADMAC with various content of comonomers were also investigated. (Monomers were studied in argon by DTA method. We should note that the difference turned to be insufficient. So, for all compounds we present the data obtained in air). The data of thermo-physical investigations are summarized in Tables 4.15, 4.16 and in Figures 4.29–4.32.

Table 4.15. Thermal characteristics of monomers and homopolymers*.

No.	Substance	DTA curve		DTG curve	Decrease of mass by 10%**	$M_{res.}$***
		Tmel. 1 (2,3)****	ETED	Decomposition interval		
Monomers						
1	GA	172 (173,175)	-	173-200, 200-275, 275-350	200	33*****
2	MAG	165 (168, 162)	-	202-270, 270-357	222	14
3	DAGA	200 (decomp.) (212,212)	-	170-293	200	9
4	DAGTFA	155 (158, 157)	-	150-222, 222-278	186	6
Homopolymers						
5	PAG	-	210-255	210-295, 295-400	235 (270)	30*****
6	PMAG	-	40-100, 178-250, 272-330	40-178, 178-272, 272-362	112 (259)	31*****
7	PDADMAC	-	75-125, 162-187, 217-248	75-162, 162-217, 217-300	117 (190)	19

Notes: * – All values in Table are presented in °C except the data on $M_{res.}$:

** – In brackets there are results corrected with consideration of water presence in structure: for homopolymers (experiments 5–7) mass was reduced by 30%;'

*** – The mass of residue after experiment, %;

**** – The melting temperature by DTA data – 1, DCT data – 2, by measuring in capillary – 3;

***** – At given residual mass foaming of analyzed sample occurs that leads to sealing and breaking of DTG curve course.

Table 4.16. Thermal characteristics of copolymers*.

No.	Substance	DTA curve Endo-thermal effect of decomposition	DTG curve Decomposition interval	Mass reduction by 10%**	M_{res}***
1	DADMAC-DAGA (87:13)	42-112, 157-220, 265-293, 317-346	42-157, 157-265, 265-317, 317-377	123 (250)	25
2	DADMAC-DAGA (90:10)	45-112, 210-215, 288-348, 382-415	45-150, 210-288, 288-382, 382-427	150 (250)	23
3	DADMAC-GA (18:82)	65-130, 205-250, 322-365, 405-430	205-227, 227-283, 283-365, 365-450	130 (260)	11
4	DADMAC-GA (34:66)	65-135, 232-255. 300-370	232-295, 295-370	135 (267)	20ʳ
5	DADMAC-MAG (43:57)	62-175, 350-405	175-235, 235-350, 350-440	150 (281)	10
6	DADMAC-MAG (24:76)	70-130, 200-240, 337-400	200-280, 280-337, 337-430	150 (280)	10
7	DADMAC-MAG (5:95)	60-140, 220-257, 355-395	220-290, 290-425	141 (273)	10

Notes: * – All values in Table are presented in °C except the data on $M_{res.}$:

** – In brackets there are results corrected with consideration of water presence in structure: for copolymers DADMAC–DAGA (experiments 1, 2) mass was reduced by 20%, DADMAC–(M)GA (experiments 3, 7) mass was reduced by 30%;

*** – The mass of residue after experiment, %;

**** – The melting temperature by DTA data – 1, DCT data – 2, by measuring in capillary – 3;

***** – At given residual mass foaming of analyzed sample occurs that leads to sealing and breaking of DTG curve course.

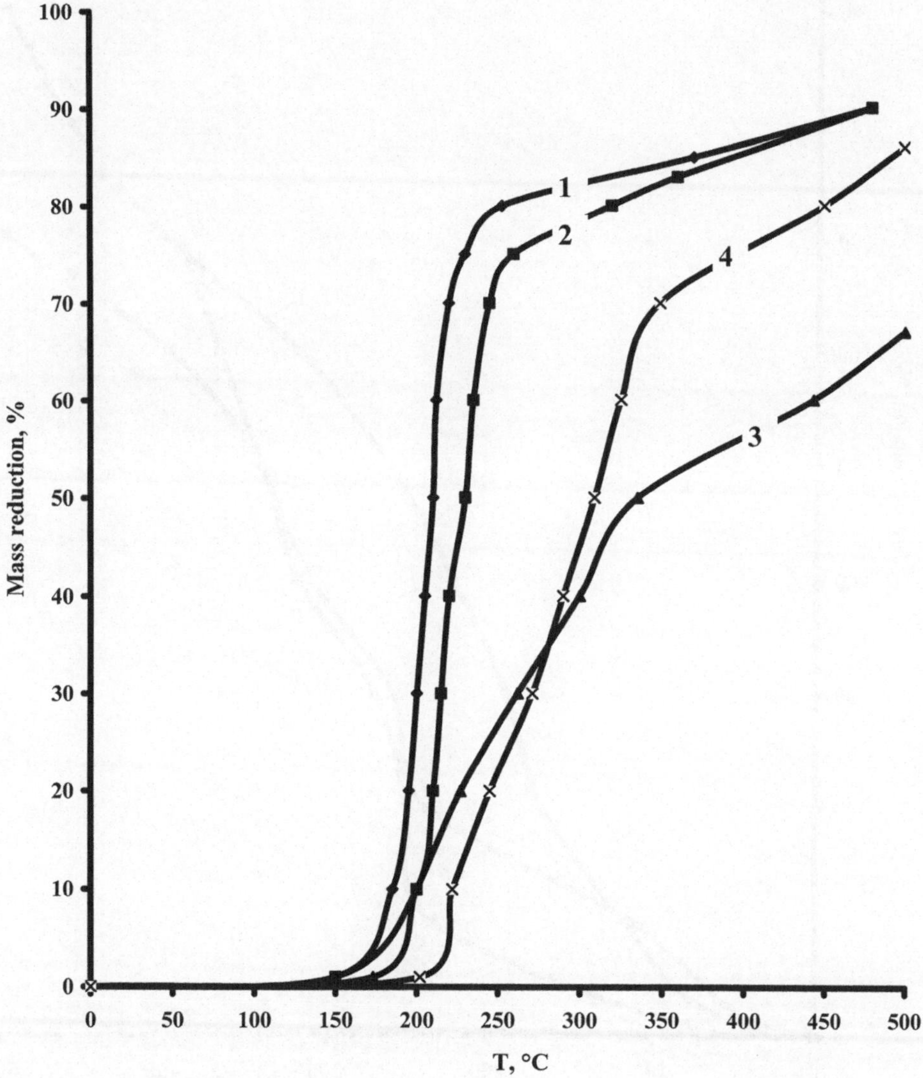

Figure 4.29. The dependence of mass change on temperature, where **1** – DAGA, **2** – DAGTFA, **3** – GA, **4** – MAG.

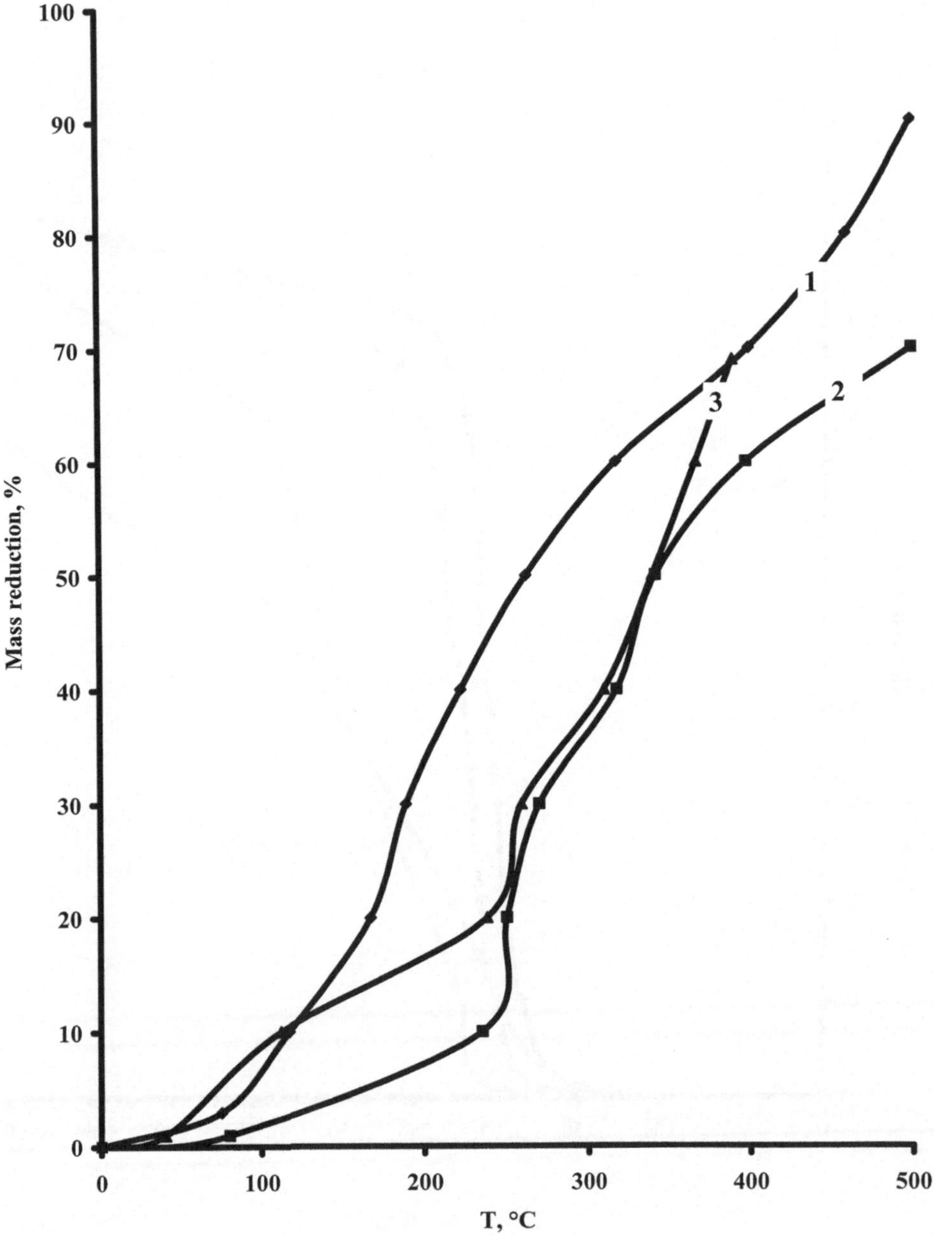

Figure 4.30. The dependence of mass change on temperature, where **1** – PDADMAC, **2** – PAG, **3** – PMAG.

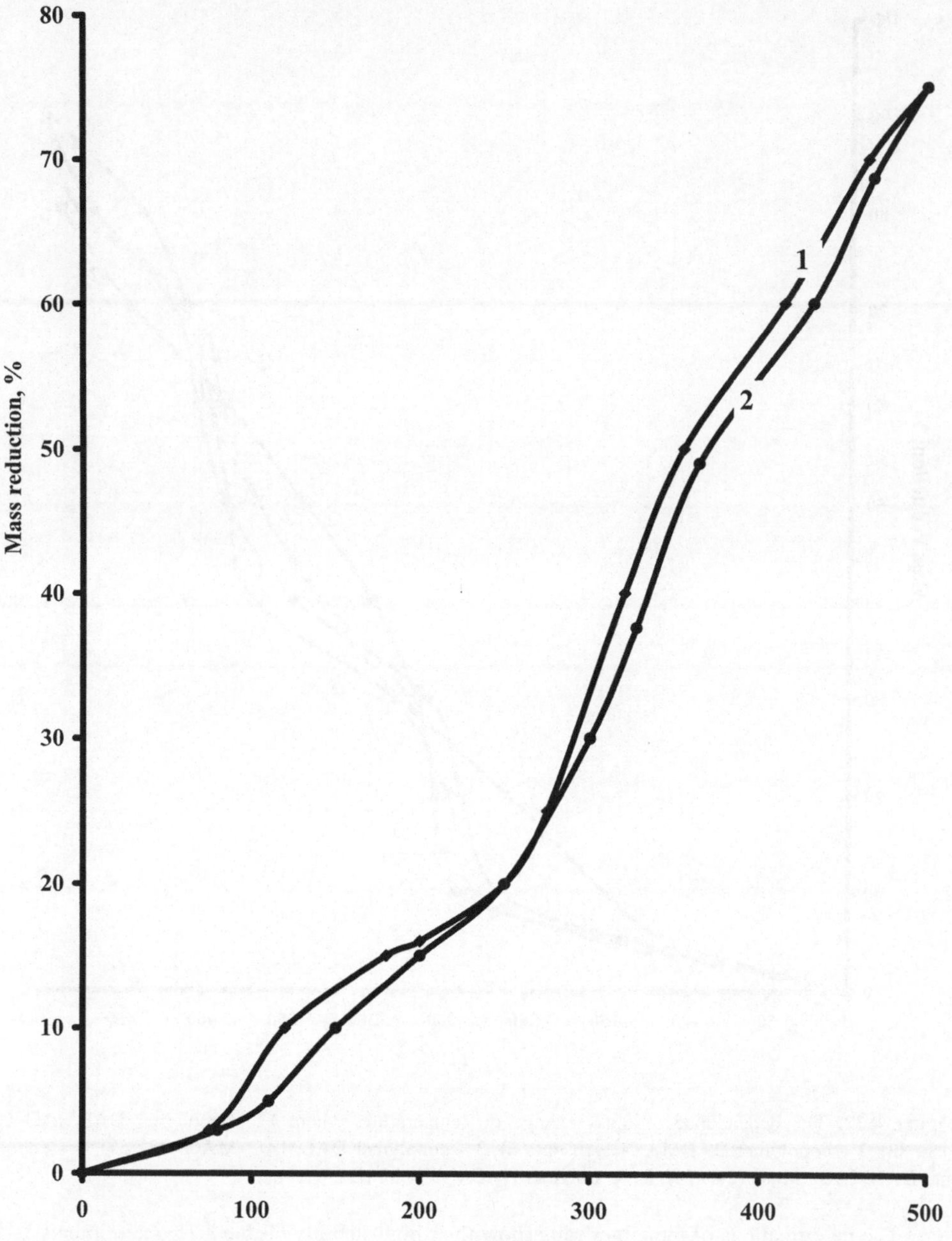

Figure 4.31. The dependence of mass change on temperature, where **1** – copolymer DADMAC–DAGA (87 : 13), **2** – copolymer DADMAC–DAGA (90 : 10).

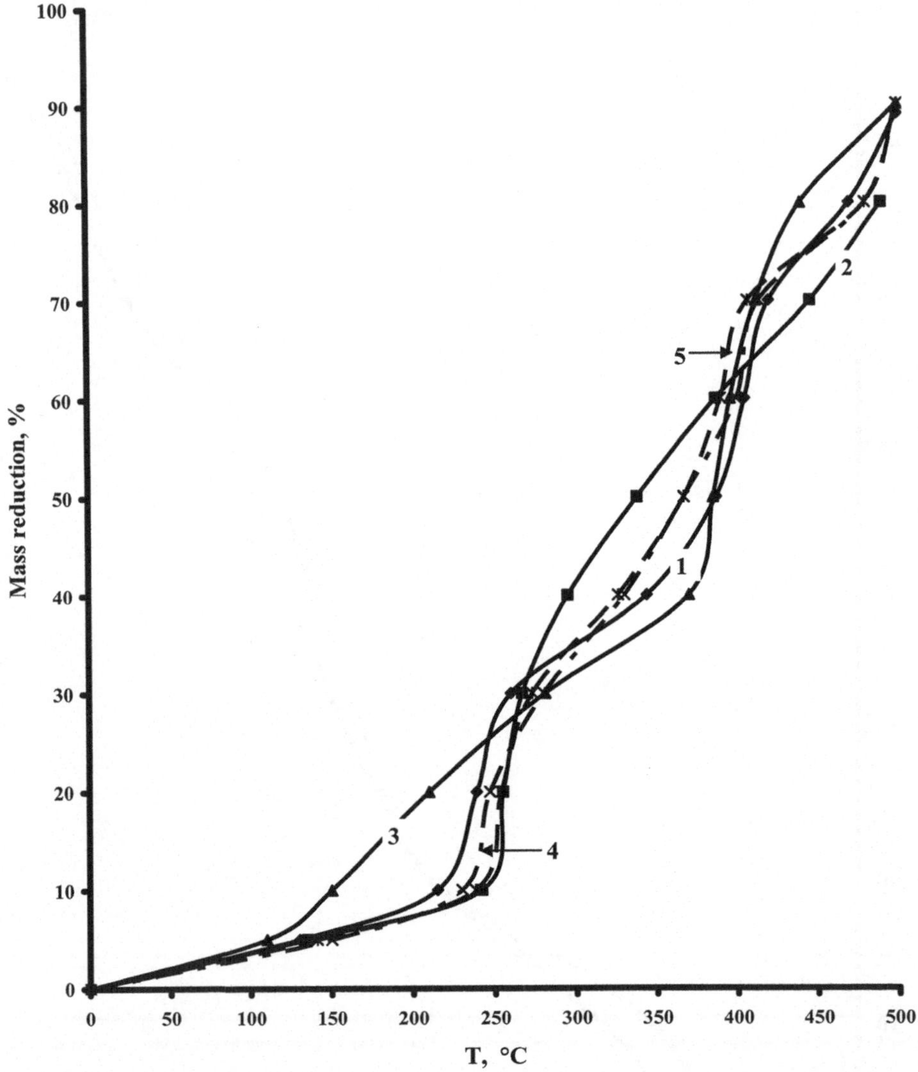

Figure 4.32. The dependence of mass change on temperature, where **1** – copolymer DADMAC–GA (18 : 82), **2** – copolymer DADMAC–GA (34 : 66), **3** – copolymer DADMAC–MAG (43 : 57), **4** – copolymer DADMAC– MAG (24 : 76), **5** – copolymer DADMAC– MAG (5 : 95).

The investigations of monomer salts show their high stability (Table 4.15, experiments 1–4). We should note, that for melting temperature both methods give good coincidence between each other and with measurements in capillary (Table 4.15). MAG turned to be more stable, mass decrease by 10% for it was observed at temperature 222°C, whereas for GA this parameter was equal to 200°C (Table 4.15, experiments 1 and 2, Figure 4.29, curves 3, 4). Fast decrease of monomers mass in the interval 230–350°C was connected obviously with the lost at the first stage of guanidine residues decomposition. As a result the further destruction proceeded with formation of the volatile products that led to foaming of investigated samples and for GA it was expressed to a greater extent. The analogous process was observed also for polymer samples and as a result of this in a number of cases we succeeded in studying of thermal stability up to 500°C not for all samples. In the case of DAG derivatives such foaming was not observed and their stability turned to be on the level of GA for DAGA (Table 4.15, experiments 3 and 4, Figure 4.29, curves 1 and 2). Fast reduction of DAGA and DAGTFA monomers mass in the interval 190–240°C is obviously connected with the fact that at the lost at the first stage of acid residues decomposition very non-stable structure of DAG is formed which is immediately oxidized with formation of volatile products.

Somewhat higher stability of DAGA in comparison with DAGTFA (Figure 4.29, curves 1 and 2) may be explained probably by higher thermal stability of structure IV characteristic for DAGA in comparison with structure V characteristic for DAGTFA. (Such structure of monomer salts including those ones in solid state was confirmed by IR spectroscopy method, see section 2.1.2, Scheme 2.5 and section 4.1).

For polymer systems somewhat surprising beginning of sample's mass lost was observed at lower temperatures than for monomer salts. This fact may be explained by the presence of water included into polymers structure (water remained also after drying for several days in vacuum drier over P_2O_5 and at temperature ~40°C). The first low-temperature interval of endothermic effect of decomposition for polymer products is connected namely with water lost (Table 4.15, experiments 5–7, Figure 4.30 and Table 4.16, Figures 4.31 and 4.32), the only exclusion is PAG (Table 4.15, experiment 5). It turned out according to DCT data that in samples of homo- and copolymers taken for investigation there was about 20% of water, i.e. such characteristic of compounds thermo stability as lost of 10% of mass required the correction of DTA data for polymer compounds. At that we should note that water in copolymers is bonded stronger than in PMAG: when studying by DCT method heating of PMAG samples up to 150°C with further cooling and again heating shows that water is completely removed out of given compound that was not observed for copolymers.

When comparing thermal stability of polymer products we should note that the most stable ones in the whole temperature interval were copolymers in comparison with homopolymers that might be easily observed by the data of Tables 4.15 and 4.16, and also in Figures 4.30–4.32. So, copolymers' mass decrease (it was mentioned that for polymer products it was 30%) was observed at 250–280°C (Table 4.16, experiments 1–7), and not at 190–260°C (Table 4.15, experiments 5–7 as for homopolymers. Intervals of endothermic effect of decomposition for all copolymers were observed at higher temperatures (please compare the same data). The only one exclusion is PAG which has comparable with copolymers temperature of mass lost (Table 4.15, experiment 5). Probably it is connected with the fact that its sample contains insufficient amount of water which may serve as destabilizing agent during heating: lost of PAG's 10% of mass was observed at 235°C in contrast to other polymer substances.

We should also note that thermal stability of copolymers with acrylate guanidine containing monomers is higher that stability of copolymers with diallylguanidine monomers. Probably, it is connected with more complex structure of copolymers DADMAC–GA and DADMAC–MAG possessing additional stabilizing intra-molecular ion bonds of variously charged units (so-called units $\mathbf{M_4}$ and $\mathbf{M_5}$). Actually, if in less stable sample of copolymer DADMAC–GA (Table 4.16, experiment 3) there are no such units, then in the other sample (Table 4.16, experiment 4) there are about 18% of them (Table 4.14, experiments 2 and 4). For samples of copolymers with MAG the stability of samples is reduced from 5 to 7 (Table 4.16), the content of units $\mathbf{M_4}$ and $\mathbf{M_5}$ in them is decreased analogously (Table 4.14, experiments 7, 4 and 2).

Thus, physical-chemical investigations confirmed the conclusions about structure of guanidine containing monomer and polymer compounds. So, it follows from the data of IR and NMR spectroscopy for example that structure of MAG is similar in both solid state and solutions (the first type of bonding with two hydrogen bonds, scheme 4.1). Whereas structure of GA is similar in solid state and in DMSO (analogous structure in DMSO for MAG), in water it has other type of hydrogen bonding (with one hydrogen bond with displacement of proton to guanidine group, Scheme 4.2). In homopolymers investigated in water by NMR method the particularities of structure characteristic for corresponding monomers are maintained. The same particularities of hydrogen bonding of GA and MAG are aspired to keep also in copolymers with DADMAC: in the case of GA this is friable ball-shaped structure with inclusion of water molecules, and in the case of MAG at the expense of strong hydrogen bonding and hydrophobic interaction this structure is more compressed. Thermal physical investigations also confirm these conclusions. Thermal stability of copolymers is higher than for homopolymers that confirm formation of intra-molecular pairs of counter-ions variously charged units of ampholytic copolymers.

We should also note that physical-chemical investigations showed that only DAGA had the second type of bonding (Scheme 4.1 and 4.6) that was not observed neither for DAGTFA, GA and MAG, nor in polymer products received on their base.

Chapter 5. Biocyde activity of polyelectrolytes

At present polyelectrolytes (PE) find an application for various branches of industry and medicine as physiologically-active substances (PAS), carrier of medical forms, immunostimulant, antimicrobial preparations, etc. [1–7].

Nowadays it is hard to find such group of materials on which micro-organisms don't exert destructive effect. Vital functions of various pathogenic microbes cause both undesirable changes of structure and functional characteristics of materials and products and realization of microbes' destructive action inside of organisms' bioplasts. The cardinal problem of modern chemistry of biocide compounds is the increase of effectiveness of antimicrobial therapy. The wide distribution of strains of pathogenic organisms of various infections stable to a lot of biocide substances and possibility of their epidemic spread make us thinking about creation of more perfect forms of biocide preparations. So, the search of means influencing on formation of stability features and providing simultaneous blocking of several factors of stability becomes necessary.

Application of biocide polymers seems to be perspective for solution of these problems. Polymers allow both including various biologically active substances differing in mechanism of action on bacterial cell into the structure of one macromolecule and providing transport of medical formula to bacterial cell-target. While they are sorbed on cell and interact with cell membranes they may influence on their functioning, in particular of barrier functions, and also to create higher local concentrations of biocide agent on the surface of bacterial cells and to form of treated surfaces the continuing polymer film providing prolonged biocide effect.

Thus, polymers due to their chemical nature open the prospects of creation on their base of biocide means revealing combined effect on bacterial cell. Insolubility in water limits possibility of use of some substances as biocide preparations. When one uses water soluble polymers as carriers this disadvantage is removed to a great extent. In complex with biocide substance water-soluble polymers acquire pronounced bactericide and bacteriostatic properties. At that harmful influence of biocide substance used for polymer marking on man's cells and tissue is significantly reduced at complete antimicrobial activity.

5.1. Biocyde properties and mechanism of biocide action of polyelectrolytes

Some examples presented below demonstrate biocide effectiveness of PE. So, carbo- and hetero-chain cationic PE reveal strong antimicrobial action. At that the highest activity was marked for polyelectrolytes containing quaternary ammonium groups in both main and side chains of macromolecules. Minimum concentrations of these polymers inhibiting for example staphylococcus growth are 50–250 $mkg{\cdot}ml^{-1}$ [8–11]. In works [12–15] they showed that diallyl cationic polyelectrolytes containing quaternary ammonium groups revealed bactericide effect on Treponema pallidum (causative agent of syphilis).

Polymer salts of antibiotics completely keeping their spectrum of antimicrobial action and providing high concentrations of antibiotics in blood and lymph of experimental animals for a long time were synthesized by D.A. Topchiev and V.A. Aleksandrova in IPCS RAS [16]. As counter-ion for quaternary nitrogen atom of polymer chain the benzylpenicillin acid was added. As initial polymers for creation of such preparations the homopolymer PDADMAC with molecular mass 25000 and copolymer of DADMAC with acrylic acid with structure 50 : 50 were taken. On the base of results of biological testing the authors showed that resulted salts exerted pronounced prolonged action and allowed decreasing of sum amount of antibiotic necessary for course of treatment in two times.

Japanese researches in works [17–19] studied polymers of quaternary ammonium compounds and biguanides and investigated their comparable antimicrobial activity in dependence on molecular mass. They revealed that polymer biguanides with molecular mass 11900 possessed significantly higher bactericide action in relation to goldish staphylococcus in comparison with monomers and polymers on the base of quaternary ammonium compounds with molecular mass 14300

Since guanidine derivatives are significantly more effective than quaternary ammonium compounds they find a wide application as physiologically active substances: medicines, antiseptics, pesticides [6, 20, 21]. Guanidine compounds are widely spread in nature. The amino-acid arginine, folic acid, numerous proteins and nucleic acids in which guanidine derivatives enter as various structure elements relate to them. Guanidine group serves as the beginning of a lot of medicines (sulgine, ismeline, faringosept) and antibiotics (streptomycin, blasticydine, mildomycine). There are guanidine derivatives among specific substances with the help of which plants are protected from micro-organisms attack [22].

The first data on biocide properties of guanidine derivatives and polymers on their base were published in patent literature [for example, 23–26]. In mentioned patents the application of analogous compounds as insecticides is described and they noticed that corresponding compounds are especially active against fungus diseases of fruit trees.

The most strong among known guanidine antiseptics are 1,6-bis-4,4-chlorophenoxybiguanidohexine ("chlorhexidine") and low-molecular polyhexamethylenebiguanidine ("vantocyl" or "cosmocyl") [21]. So, for example chlorhexidine was used as disinfectant mean as salts (hydrochloride, acetate, gluconate) and was widely recommended until recently in the form of solutions, ointments, powders as effective disinfectant mean in surgery for fight with hospital-acquired infections, for treatment of skin diseases and community aims. However, we should note that these substances are prepared by complex 4 stage technological scheme. Moreover chlorocyan is the initial raw material for their synthesis that is why the technological process if expensive and dangerous.

In Russia (as it was mentioned in the Introduction in IPCS RAS) the process of production of guanidine antiseptic – polyhexamethylene guanidine hydrochloride (PGMG) ("polysept") from hexamethylene diamine and guanidine was developed [6]:

$$nH_2N\text{–}(CH_2)_6\text{–}NH_2 + nH_2N\text{–}\underset{\underset{NH\cdot HCl}{||}}{C}\text{–}NH_2 \rightarrow [\sim (CH_2)_6\text{–}NH\text{–}\underset{\underset{NH\cdot HCl}{||}}{C}\text{–}NH \sim]_n + 2NH_3$$

Since three amino-groups of guanidinehydrochloride have various reaction abilities, then molecular mass and structure of "polysept" can be controlled by changing of reaction conditions and content of hexamethylene diamine in initial mixture [6, 27]. So, at comparatively low temperatures for polycondensation process (120–130°C) predominantly two amino-groups react with hexamethylene diamine and form well soluble linear oligomers with molecular masses $1{,}7\text{-}12{,}5\times10^3$. When the amount of hexamethylene diamine in reaction mixture was increased up to more than 1 mole of guanidinehydrochloride and temperature was increased up to 180–200°C the third amino group might react and branched polymer was formed with molecular mass $20\text{–}43\times10^3$.

Various salts of PGMG (phosphate, gluconate, sorbate, fluoride, sulfate, nitrate, silicate, acetate, stearate, oleate, fumarate, succinate, adipinate, sebacinate) were obtained under the action of various acids or their salts on the base and carbonate of PGMG [6]. Among listed polymer salts in addition to hydrochloride the phosphate and gluconate have the highest practical importance. According to data of work [6] hydrochloride solutions with concentration 0,1–0,05 mass % cause the death of gram-positive and gram-negative microorganisms: c. Duphtheretiae, St.aureus, and also St.aibus and St. faekalis, S.typhi, Shigella Sonnae, Flexneri, E.Coli, Salmonella th.murum, Proteus Vulgarus, Ps.aeruginosa for 5–25 minutes.

Taking into account all said above and presented examples the development of novel biocide preparations is undoubtedly the issue of the day. Analogously to present in book [28] criteria used in the approach to medicines and essential part of which is biocide preparations these key moments may be considered in relation to antimicrobial means also called as biocide and bactericide.

1. Biocide substance should possess high activity, selectivity and continuity of its action;

2. It should be non-toxic for man and animals and shouldn't initiate any side effects;
3. It should be pure and has high stability during storage;
4. Prime cost of its production shouldn't be very high;
5. It should be available, and its profitability on biocide preparations market should high.

In correct combination all these factors will be determine the life-time of given preparation among biocide preparations functioning analogously.

The strategy and tactics of creation of novel biocide substances may be based in the following key moments [28]:

1. the principle of chemical modifying of structure of known synthetic and nature biocide substances. At that one proposes that biological activity of novel compound will be high;
2. the principle of introduction of biocide group of known substance into the molecule of novel substance;
3. the principle of molecular modelling in which the spatial structure of biocide and bioreceptor part of its capture are taken into account;
4. the strategy of pro-biocide when active element is transported to the place of action by some combined molecule from which in concrete place the biocide substance is evolved;
5. the conception of anti-metabolite, i.e. substances which substitute metabolite in natural bioreactions and disturb their normal proceeding, since they are unable to substitute the nature object in corpora;
6. the principle of application of polymer matrix or polymer substance which first of all leads to prolonged action and control of necessary biocide.

The mechanism of polyelectrolytes biocide action. Taking into account that by proper physiological activity of polymer one understands the activity which is connected with polymer state and is not characteristic for low-molecular analogous or monomers [4] the mechanisms of proper physiological activity display may involve the both such important part as physical effects connected with high mass, osmotic pressure, conformational rebuildings, etc., and also may be connected with inter-molecular interactions with organism's biopolymers. Many of organism's biopolymers are polyanions (proteins, nucleic acids, a number of polysaccharides) and bio-membranes also have sum negative charge. The interactions between oppositely charged polyelectrolytes proceed cooperatively and resulted polycomplexes are stable enough. It is known that for such interactions the charge density and molecular mass [4, 5, 17–19] are of the highest importance. And if we speak about biocide properties then the knowledge of mechanism of action plays important role.

The consequence of elemental acts of lethal action of PE on bacterial cells may be presented as follows [5]:

1) adsorption of poly-cation on the surface of bacterial cell;
2) diffusion through cell wall;
3) bonding with cytoplasmic membrane;
4) destruction or destabilization of cytoplasmic membrane;
5) evolving of cytoplasmic components out of cell;
6) cell destruction.

This consequence relates to polycations first of all since bio-membranes have negative sum charge, however negatively charged in the whole cell membranes have isolated poly-cationic regions on which the poly-anions may be sorbed [29]. Since the main stages of influence of biocide preparation are connected with the structure of cell membrane of bacteria we want to present briefly the main moments of its organization [5, 21].

Cell membrane functionally separates the internal medium of cell from environment. It provides cell by mechanical stability and allows organism to grow and to be divided, determines the form and sizes of bacterium. The sizes of cell may be changed in dependence on degree of change of physical and chemical conditions of cell surrounding, especially the structure of nutrient medium. Cell membrane includes two main components: cell wall and cytoplasmic membrane.

Cell wall has on its surface specific anti-gene components, receptors and other additional elements. In spite of its mechanical stability and relatively compact structure cell wall provides small molecules with entrance and exit from cell and also creates the ways for macromolecules transport. The cell also has places of macromolecular material moving out through which the extracellular ferments and exotoxins are evolved. The structure of bacterial cell walls is widely varied in dependence on the type. It was studied and described in details. The most noticeable differences were revealed between walls of gram-positive and gram-negative bacteria. The wall of gram-positive bacteria represents thick rigid structure which thickness is about 15 up to 100nm (usually it is 15–50 nm). The main component (from 60 up to 90%) by mass of cell wall is peptidoglycan also called as mucopeptide or murean. The structure of peptidoglycan is varied in dependence on bacteria type.

The main chains of peptidoglycan are built of alternate residues of N-acetylglucoseamine and N-acetylmuramic acid. These chains are cross-linked with each other by peptide bridges. Peptidoglycan forms multi-layered structure with lateral cross-links in both inside of the layer and between layers. Peptidoglycan's function is keeping the form of bacterial cells. It is hydrolyzed by some ferments, for example, lysozyme and linear water-soluble short polymer chains are formed at that.

Together with peptidoglycan in cellular net of gram-positive bacteria there are also hydrophilic acid polysaccharides poly(D-ribitol-5-phosphate) or substituted glycerol-3-phosphate also called as teichoic acids which bonded with peptidoglycans covalently. In some bacteria the teichoic acid is substituted by poly(N-acetylglucoseamin-1-phosphate) called as teichuronic acid. Presence of these polymer acids imparts the cell wall of gram-positive bacteria by negative charge. Ability to retain the gentian violet dye under dying according to Gram to a significant degree depends on large content of peptidoglycan (if it was extracted in pure form it easily absorbed this dye), on the presence of teichoic acid characteristic only to gram-positive bacteria and also on special structure of wall skeleton in which pores after treatment by alcohol are converged trapping the dye.

Cellular membranes of gram-negative bacteria with thickness equal to 10–15 nm are more complex. They together with peptidoglycan layer also have so-called external membrane which contains phospholipids, liposaccharides, lipoproteins and proteins and represent the double lipid layer. Phospholipids in it are distributed asymmetrically. The external part of bio-layer contains liposaccharides and some proteins, whereas internal one contains mainly only phospholipids and small amount of proteins. Lipopolysaccharides contain hydrophilic polysaccharide part which spread into external space and hydrophobic segment called as lipid A which is immersed into hydrophilic region. peptidoglycan is presented by one layer situated under liposaccharides. This internal layer is a peculiar bag of peptidoglycan. Due to high content of lipids cellular net of gram-negative bacteria is less permeable and its pores after treatment by alcohol remain wide enough that leads to removal of gentian violet dye during dying according to Gram. Lipopolysaccharides provide cells surfaces of gram-negative bacteria by negative charge.

One more important particularity of lipopolysaccharides is in the fact that they provide bonding with external membrane of membrane proteins and also divalent cations Ca^{2+}, Mg^{2+}, etc., which provide additional cross-linking by non-covalent ion bridges. Treatment of gram-negative bacteria' cells by ethylendiamintetraacetic acid chelating divalent cations leads to destruction of external membrane.

External membrane of gram-negative bacteria fulfills definite barrier functions and imparts to these organisms high level of stability to a lot of antibiotics, detergents and other substances. Essential differences in ultra-thin structure of cell wall of gram-positive and gram-negative microorganisms in a number of cases explain their various sensitivity t β- lactam and other antibiotics influencing n synthesis of cellular wall.

Cytoplasm membrane is the main barrier separating cell from environment and its safety is the necessary condition of existence of cell as a single whole. It provides constancy of intracellular structure independently on changes in the environment. Cytoplasm membrane consists mainly of phospholipids and proteins and has complex structure organization which base is bi-layered organization of phospholipids. In bi-layer hydrophilic parts of molecules of polar lipids are turned into water phase and hydrophobic (residues of fatty acids) form the numbers of hydrocarbon chains parallel to each other which are retained by hydrophobic interactions. At that the chains of fat acids of phospholipid molecules are elongated inside of bi-layer and polar ends are oriented outside. Thickness of lipid bio-layer is varied in the limits from 6 up to 10nm.

Molecules of proteins and phospholipids entering into bi-layer structure at the expense of heat motion at physiological temperatures freely move as along the layers, so between them. At temperature decrease when diffusion at the expense of heat motion is hindered various structures are formed inside of the layer. Such as:

1. lipid–lipid clusters and due to it appearance phase separation is observed;
2. protein–lipid clusters;
3. domains consisting of intra-membrane components resulted from their bonding with high-molecular compounds situated on membranes surfaces;
4. protein–lipid domains differing from other regions of membrane in structure, etc..

At the expense of formation of such structures micro-heterogeneity is appeared in membrane playing an important role in membranes functioning, in change of their permeability in relation to various low-molecular substances including active transport, electron transport, oxidative phosphorylation, fermentative synthesis of cellular wall, etc..

Ion of metals Mg^{2+}, Ca^{2+}, non-polar lipids in particular sterols significantly influence on stability and permeability of model and biological membranes. In the presence of Mg^{2+} and Ca^{2+} the bridges between neighboring phospholipids are formed, as a result packing becomes more compact and membrane permeability is decreased. Such effect is observed under increase of fat acids and cholesterol residues content. Water forming 20-35% of membrane fractions plays a great role in structure and functioning of membranes.

Structural stabilization of membranes is determined by two types of non-covalent interactions – hydrophilic and hydrophobic. There also some other interactions – electrostatic and dipole-dipole between ion pairs, hydrogen bonds which make their contribution into structure stabilization. Additional stabilization is realized at the expense of association of glycolipid complexes with polar and non-polar parts of lipid molecules.

When studying the influence if various factors on the level of anti-microbe activity of cation polyelectrolytes they showed in [5, 30] that their activity was increased with the rise of number of ionogenic groups in macromolecules [5], molecular mass and character of ionogenic groups distribution along the chain didn't significantly influence on the level of anti-microbe activity. The use of polyelectrolytes with luminescence mark [31] when investigating interactions of polymers with erythrocytes and bacterial cells showed that polymer was quickly bonded by cell wall and cytoplasm membrane and then penetrated into cytoplasm and cell nucleus. At that the permeability of cell membrane was increased for both low-molecular [5, 30, 31] and high-molecular substances [32]. The rise of polyelectrolytes concentration up to 50-100 $mkg \cdot ml^{-1}$ and higher leads to intensive damage of cellular membrane determined by evolving of proteins and nucleic acids out of cells [5, 30–33].

The key moment in the mechanism of action of cation polyelectrolytes on biological membranes is electrostatic interaction with negatively charged phospholipids and proteins localized on it. The consequence of this is neutralization of membrane charge and cell in the whole and also the change of ratio of hydrophobic and electrostatic interactions stabilizing the membrane.

In work [34] they studied the behavior of polyelectrolytes in biological environment. For investigation of cells interaction with polyelectrolytes authors used model systems – bi-layered lipid vesicles. Behavior of the system polycation–vesicle was studied in dependence on structure and liner density of polycation charge, membrane phase state, content of charged lipid in vesicles and their sizes, and also on ion force of solution. They showed that given interaction might be accompanied by lateral aggregation of lipids, sharp acceleration of trans-membrane migration of lipid molecules (flipflopping), hindering of lipids exchange between vesicles, embedding of adsorbed macromolecules into vesicle membrane, and also by aggregation and destruction of vesicles. Polycations adsorbed on membrane only at the expense of electrostatic interactions may be quantitatively displaced to the solution at the increase of ion force of solution or at addition of excess of polyanion-competitor. This leads to the reduction of both initial distributions of lipids in membrane and inter–vesicle lipids exchange. Hydrophobization of the main chain of polycation or modifying of chain by side hydrophobic radicals provide stability to polycation–vesicle complex in water-salt solutions and in the presence of essential excesses of negatively charged polyions at the expense of embedding of hydrophobic fragments of polycation into hydrophobic part of vesicle membrane.

The obtained results are of great interest from the point of prognosis of possible consequences of contacts between polyelectrolytes and biocide substances on their base with cellular surface. Thus, destabilization of the membrane may lead to the change of localization and phospholipid environment of ferments connected with membrane that naturally influences on their activity. Elucidation of influence of cation polyelectrolytes on bacterial ferments causing stability of bacterial cells to antibiotics represents undoubted interest.

It turned out [35] that cation polyelectrolytes interact also with bacterial ferments. The inhibiting effect of quaternary ammonium salts of polydiethylamine ethylmethacrylate and polydimethylamine isopropylmethacrylate on "aggression" ferments (factors of pathogenicity) of golden staphylococcus – plasmo-coagulase and hyaluronidase causing its pathogenicity was found. These polymers also inhibit the ability of staphylococcus α-toxin to hemolyze rabbit's erythrocyte. Cationic polyelectrolytes reveal inhibiting action also in relation to bacterial ferments inactivating antibiotics, penicillinase hydrolyzing amide bond of β-lactam cycle of penicillins and transferring penicillins into inactive penicillin acids. Property of cation polyelectrolytes to suppress activity of bacterial ferments inactivating antibiotics and also to increase permeability of cellular wall and cytoplasm membrane may promote strengthening of antibiotics action in relation to resistant (stable to antibiotics action) cultures of bacteria since at that the conditions are created facilitating the antibiotics goal achievement in cell. That is why cation polyelectrolytes are of great interest as novel anti-microbe substances and as membrane-acting biologically active polymer-carriers for low-molecular biocide substances.

Thus, according to available at present data the mechanism of biocide action of cationic polymer biocide substances, in particular guanidine-containing ones is as follows. Since microorganisms usually possess negative sum electric charge they are substantive to cation of bactericide preparation which being contacting with microorganism is adsorbed on cellular membrane, causes its lysis and penetrate inside of cell. Being inside of it the preparation exerts blocking effect on biological activity of ferments, hinders replication of nucleic acids and oppresses respiratory system. This complex of effects in its totality leads to microorganism death.

In comparison with other cationic polyelectrolytes application of guanidine-containing biocide polymers has one advantage. When using synthetic biocide polymers one should take into account their bio-degeneration in living organism. In the case of use of non-bio-degradable synthetic polymers their molecular mass is of great importance since polymers with molecular mass higher than 50000 can't be removed by kidneys and are accumulated in renal tubules causing expressed toxic effects. In this relation the biocide effect of guanidine compounds may turn to be physiological, since in living organism there are fermentative systems able to initiate degradation of these compounds preventing their accumulation [6]. However, as it was mentioned above for each novel guanidine containing compound given fact requires separate examination.

In relation to possible action of guanidine fragment fulfilling in our case the biocide function we may express our following view. The amino-acid arginine contains the guanidine residue and under proceeding of various processes in cell the guanidine fragments of our polymers may substitute arginine causing various damages. For example, the following forces play the most essential role in stabilization of tertiary structures of proteins: covalent bonds, ion bonds, hydrogen bonds, hydrophobic interactions and Van der Waals forces. And ionic bonds are appeared between ionogenic radicals of amino-acids. Aspartic and glutamic acids from one hand and lysine or arginine from the other participate in their formation more often. Guanidine residues may also damage nature stabilization of tertiary structure at the expense of hydrogen bonds which are appeared between polar groups of side radicals and also between side radicals and peptide groups. Polymer chains of "strange" polymer may also participate in damage of hydrophobic interaction. Moreover, one of the factors causing denaturation of proteins, i.e. destruction of secondary, tertiary and quaternary structures characteristic for proteins in nature is addition of various chemical reagents including guanidine. In the case of copolymers of DADMAC and (M)AG this compound is delivered to the cell by polymer matrix and bonded with the help of the last one destructing its integrity that allows guanidine molecules to penetrate inside of cell at the expense of passive transport which is determined only by difference of concentrations or field direction.

Analogous interference may occur in processes in which nucleic purine base guanine participates having in its structure guanidine fragment. The main reason of polynucleotide chains association is the formation of hydrogen bonds between nucleic bases (guanine forms pair with cytosine, pyrimidine base). Substitution of guanine by guanidine fragment may cause destructions in such important processes as genetic information transfer, nucleic acids synthesis (DNA, RNA) and some others.

5.2. Biocide and toxicological properties of synthesized novel guanidine containing polymer products

All said above testifies to perspectiveness and principle possibility of use of guanidine-containing polymer substances synthesized by us as biocide preparations. We should note, that these polymers meet the number of requirements which are made of modern preparations of such type: good solubility in water and physiological solution (1% solutions of polymers have pH = 6,5–7,0); the solutions are colorless, don't have smell, don't cause destruction of treated materials and polymer nature of these compounds promotes the absence of inhalation toxicity and formation on treated surfaces of continuing polymer film providing prolonged biocide effect.

Investigations of bactericide activity and toxicity of synthesized homo- and copolymers carried out in laboratory of polyelectrolytes of IPCS RAS together with N.F. Gamalei's Institute of micro-biology and epidemiology RAMS, bacteriological laboratory of State sanitation-epidemiological supervision of Kabardino-Balkaria republic (KBR) and pharmaceutical association "El'farmi" (KBR, Nal'chik) showed that these preparations were very active and posses biocide action in relation to gram-positive (St.Aureus) and gram-negative (E.coli) microorganisms and that a lot of samples had low toxicity (Table 5.1). The technique of determination of bactericide activity and toxicity is presented in Appendix.

Table 5.1. The data on biocidicity and toxicity of polymer derivatives of GA, MAG and DAGA and some model polymers*.

No.	Compound	**$M_1 : M_2$****	**M_3****	**M_4*****	**M_5*****	**I_t**	E. coli r	St.aureus****	MSC×10^3*****
1	PAA	-	-	-	-	87,1	---	---	-
2	PAG	0:100	-	-	-	67,6	---	-++	8,5
3	CP with GA	18:82	10	-	-	75,2	--+	+++	1,9
4	CP with GA	36:64	14	18	18	98,8	-++	+++	2,0
5	CP with GA	75:25	3	12	12	122,4	+++	+++	2,3
6	PMAA	-	-	-	-	93,2	---	---	-
7	PMAA + PMAG	-	-	-	-	76,1	---	-++	9,8
8	PMAG	0:100	-	-	-	104,6	---	-++	9,2
9	CP with MAG	11:89	11	-	-	88,7	-++	-++	2,2
10	CP with MAG	28:72	17	16	16	101,5	-++	+++	2,7
11	CP with MAG	34:66	11	19	19	119,1	+++	+++	1,4
12	CP with MAG	43:57	6	26	26	131,3	-++	+++	1,9
13	CP with MAG	61:39	3	15	15	143,4	-++	+++	3,2
14	PDADMAC	100:0	-	-	-	29,7	--+	---	-
15	PHMG HC	-	-	-	-	27,5	-++	-++	-
16	CP with DAGA	85:15	-	-	-	38,6	+++	+++	0,4
17	CP with DAGA	95:5	-	-	-	33,5	+++	+++	1,5

Notes:* – the samples of copolymers DADMAC with GA (experiments 3–5), copolymers DADMAC with MAG (experiments 9–13), PAA (experiment 1), PAG (experiment 2), PMAA (experiment 6), PMAA modified by guanidine (experiment 7), PAG (experiment 8), PDADMAC (experiment 14), PHMG HC (experiment 15), copolymers DADMAC with DAGA. Methods of investigations of biocide and toxic properties are presented in Appendix; **I_t** – index of toxicity;

** – according to ^{1}H NMR spectroscopy data;

*** – according to ^{1}H NMR spectroscopy and elemental analysis data;

**** – Escherichia coli is colon bacillus, representative of gram-negative bacteria and Staphylococcus Aureus 906 is golden staphylococcus, representative of gram-positive bacteria.

(+++) continuous lysis of bacterial cell, hinders growth of given culture completely;

(--+) partial lysis of bacterial cell, the zones of growth suppression are observed after 48 hours;

(--+) partial lysis of bacterial cell, the zones of growth suppression are observed after 72 hours;

(---) inactive;

***** – minimum suppression concentration, mass %.

On the base of obtained data we may conclude the following. Polyacrylic acids (Table 5.1, experiments 1 and 6), and also homopolymers PAG and PMAG (Table 5.1, experiments 2 and 8) possess low toxicity, and polymer derivatives of acrylic acid posses the lowest toxicity. This fact may be explained probably by additional hydrophobic interaction of polymers on the base of MAG with cellular wall. The same regularity is observed for copolymers (Table 5.1, experiments 3–4 and 9–13): copolymers with GA are less toxic, at that toxicity of samples is the higher, the higher the content of DADMAC units in copolymers (Figure 5.1). The presented Figure needs the following explanation. Since the criterion of non-toxicity according to technique of investigation (see Appendix) is the value of index of toxicity in the interval from 60 up to 120 (Table 5.1, column 7), than for clearness during comparing the low level equal to 60 was subtracted from the value of this index.

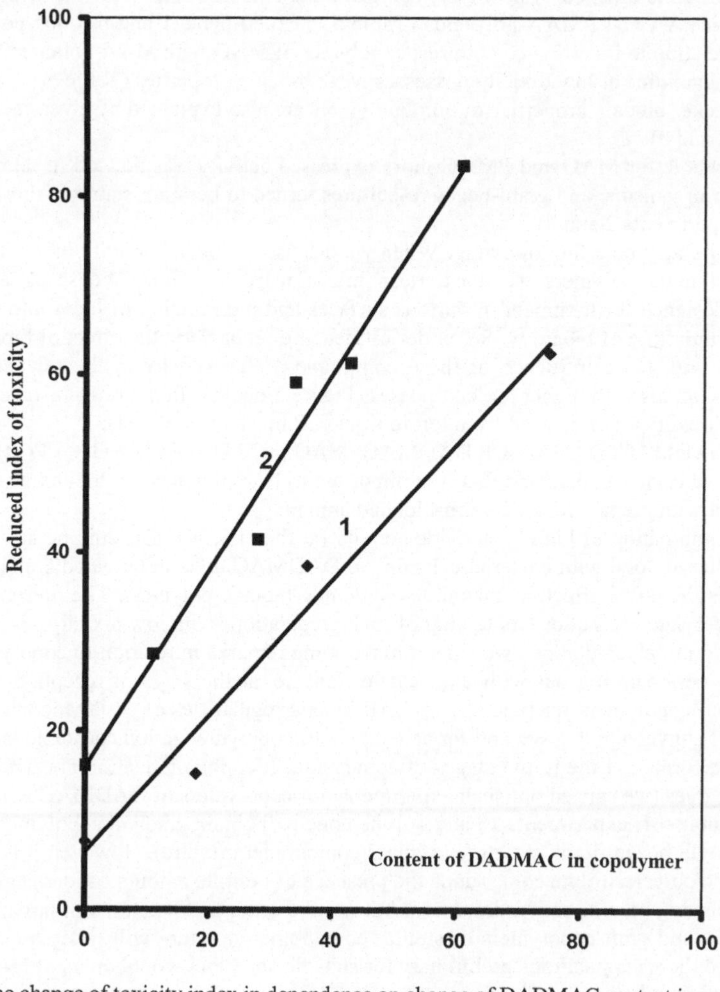

Figure 5.1. The change of toxicity index in dependence on change of DADMAC content in copolymers with GA (curve **1**) and MAG (curve **2**).

This result was expected for polyamphylic copolymers if one took into account the fact that derivative of acrylic acids had low toxicity, whereas PDADMAC possessed high toxicity (Table 5.1, experiment 14) due to its polycation nature. High toxicity was revealed also by samples of copolymers of DADMAC with DAGA however it was lower than for PDADMAC (Table 5.1, experiments 16, 17 and14). The other polycation taken for comparison polyhexamethyleneguanidine hydrochloride also showed high toxicity (Table 5.1, experiment 15). So, we may conclude that when select the copolymer as biocide preparation one should take into account the copolymer structure.

As it is obvious from data of Table 5.1 the synthesized guanidine-containing preparations reveal bactericide activity in relation to studied cellular structures and for copolymers the most expressed biocide activity was observed (the highest biocidity was revealed by copolymers of DADMAC with DAGA, Table 5.1, experiments 16 and 17). As it is known, it is connected with the fact that polyelectrolytes in water medium form electrically charged groups, fixed along the lengthy polymer chain and able to "multipoint" (cooperative) interactions with bacterial cell and transport function first of all is fulfilled by positively charged units of DADMAC. This fact is also confirmed by the fact that PAA and PMAA polyacids (Table 5.1, experiments 1 and 6) don't possess biocide properties in relation to investigated cultures; copolymer of MAG with MAA obtained by modifying of PMAA by guanidine at low toxicity possesses weak biocide properties (Table 5.1, experiment 7). Essentially weaker biocide properties of homopolymers are also explained by given facts (Table 5.1, experiments 2 and 8).

On the whole for MAG and PMAG more expressed activity was noticed in relation to gram-positive micro-organisms, and gram-negative-cultures turned to be more stable to these compounds (Table 5.1, experiments 2 and 8).

We should also note the following. When we discuss biological activity of synthesized by us guanidine-containing polymers we don't mean medicine preparations. We speak about biocide preparations designed for treatment of various surfaces and introduction of them into various compositions for imparting of biocidity. So, under influencing on bacteria the effect of biocidity will be supplemented with toxic influence of these compounds. That is why in the raw of synthezised polymers the samples with higher toxicity possess higher biocidity. In the raw presented below both toxicity and biocidity are decreased from left to right (CP means copolymer):

CP DADMAC–DAGA > CP DADMAC–MAG > CP DADMAC–GA > P(M)AG

At that we clearly understand that it is important to take into account the toxicity index in the case of contact with preparations of warm-blooded animal.

Thus, combination of high bactericide activity (at the expense of guanidine groups presence) with high ability to bond with bacterial cells due to DADMAC units in obtained copolymers allows us to synthesis the novel effective guanidine-containing biocide polymers. The presence in copolymer chain of acrylate units allows obtaining of such preparations with low toxicity.

In addition to all said above we should make some remarks in relation to copolymers production under process carrying out with high conversion, i.e. at the stage of reception of significant amount of such copolymers for practical use. All kinetic regularities of copolymerization were obtained at small conversion degrees and for calculation of copolymerization constants the copolymers were extracted formed at the initial stages of conversion. If copolymerization is carried out at high conversion degrees (we carried out such experiments for copolymers of DADMAC with DAGA and with MAG, Table 5.1, experiments 15 and 16) then one will obtain gradient copolymers (in the case of MAG they will be similar in structure to initial comonomer mixture). However, it is known [4, 5] that for biocide polyelectrolyte compounds the presence of definite amount of ionogenic groups and substituents playing biocide role is important but not the order and positions in polymer chain, i.e. statistic, block- and graft-copolymers of similar comonomer structure will possess close biocidity. (We confirmed these suppositions preliminary for statistic and block-copolymers pf DADMAC with DAGA: copolymers having similar composition possessed close biocide parameters. These works are continued now.) So, copolymer composition may be controlled at the expense of initial composition of comonomer mixture, and yield and conversion may be regulated by the ratio monomer : initiator.

Apart from the data mentioned above we found also biocide action of novel guanidine containing copolymers in relation to man's pathogenic fungal microflora, in particular Candida alb. The most active in this relation are the copolymers DADMAC : DAGA (85 : 15) (MSC 0,003%) and DADMAC : MAG (63 : 37) (MSC 0,001%).

In the laboratory of phytopathology of Kabardino-Balkarian representation of All-Russian Institute of plants protection some of synthesized samples were tested on fungicidal activity against solid brand of winter wheat (Table 5.2). The technique of fungicidal activity estimation is presented in Appendix.

Table 5.2. Effectiveness of novel guanidine-containing preparations against solid brand of winter wheat.

No.	Preparation	Concentration, %	The decrease of lesions, %.
1	PMAG	0.075	40
2	DADMAC-DAGA	0.075	70
3	DADMAC-MAG	0.075	85
4	Polycarbacine	0.075	20

Carried out investigations on estimation of fungicidal activity of PMAG and copolymers DADMAC–DAGA showed that treatment of plants by synthesized preparations (Table 5.2, experiments 1–3) in concentration 0,075% reduced the lesions of winter wheat by solid brand by 40–80%, i.e. it was more effective than known preparation polycarbacine.

References

1. R.V. Petrov, V.A. Kabanov, R.M. Khaitov, *Immunologiya*, No.1, 5 (1986) (*in Russian*).

2. V.A. Kabanov, *Macromol.Chem., Macromol. Symp.*, **1**, 101 (1986).

3. R.V. Petrov, V.A. Kabanov, R.M. Khaitov, A.V. Nekrasov, R.I. Ataullakhanov, *Allergy & Clinical Immunology*, **15**, 56 (2003).

4. N.A. Plate, A.E. Vasil'ev, *physiologically active polymers*, Moscow: Khimiya (1986) (*in Russian*).

5. G.E. Afinogenov, E.F. Panarin, *Antimicrobial polymers*, Saint-Petersburg: Gippokrat (1993) (*in Russian*).

6. P.A. Gembitzky, I.I. Vointzeva, *Polymer biocide preparation polyhexamethylenguanidine*, Zaporozhye: Polygraf (1998) (*in Russian*).

7. Ya. Kalal, *Vysokomol. Soed.*, **21A**, 2447 (1979) (*in Russian*).

8. M. Ghosh, *Polymer Material Sci. Eng.*, ACS., **55**, 755 (1986).

9. M. Ghosh, *Polymer News*, **13**, 71 (1988).

10. C.M. Samour, *Polymer drugs.*, Chemtech (1978).

11. A. Rembaun, E. Selegny, *Polyelectrolytes and Their Applications*, 187-195, 131-144, 163-174 (1975).

12. N.A. Zaikina, *Antibiotiki*, No.22, 327 (1977) (*in Russian*).

13. M.V. Milich, D.L. Fedorova, D.A. Topchiev, *Vestnik dermatologii*, No.9, 28 (1988) (*in Russian*).

14. M.V. Milich, D.L. Fedorova, D.A. Topchiev, *Vestnik dermatologii*, No.4, 37 (1988) (*in Russian*).

15. M.V. Milich, D.L. Fedorova, D.A. Topchiev, *Vestnik dermatologii*, No.5, 25 (1988) (*in Russian*).

16. V.A. Aleksandrova, V.A. Zlobina, G.A. Dmitriev, T.I. Milonova, D.L. Fedorova, D.A. Topchiev, *Khim. Farm. Zh.*, No.5, 18 (1994) (*in Russian*).

17. T. Ikeda, H. Yamaguchi, S. Tazuke, *Antimicrob. Agents Chemother.*, **26**, 139 (1984).

18. T. Ikeda, S. Tazuke, Y. Suzuke, *Macromol. Chem.*, **185**, 869 (1984).

19. T. Ikeda, S. Tazuke, *Polymer. Prep.*, **26**, 226 (1985).

20. *Chemical Encyclopedia*, ed. by I.L. Knunyantz, Moscow (1988) (*in Russian*).

21. T.J. Franklin, G.A. Snow, *Biochemistry of Antimicrobial Action.*, London: Chapman and Hall (1981).

22. T.J. Franklin, G.A. Snow, *Phytochemistry*, **48**, No.3, 465 (1970).

23. Patent 789429 France (1959).

24. Patent 2,867,562 USA (1959).

25. Patent 1114155 Great Britain (1960).

26. Patent 339076 Sweden (1971).

27. A.R. Khokhlov, S.A. Pavlova, G.L. Timofeeva, *J. Polymer.*, **35**, No.8, 1769 (1994).

28. A.T. Soldatenkov, N.M. Kolyadina, I.V. Shendrik, *The basis of organic chemistry of medical preparations*, Moscow: Khimiya (2001) (*in Russian*).

29. J. Pitha, *In: Anionic Polymeric Drugs*/ L.G. Donaruma, R.M.Ottenbrite, O.Vogel eds. Wiley - Intersci.Publ., N.Y. (1978).

30. E.F. Panarin, *26 Microsymposium on Macromolecules Polymers in medicine and Biology.* Prague (1984).

31. N.A. Plate, A.E. Vasil'ev, *Vysokomol. Soed.*, **24A**, No.4, 675 (1982) (*in Russian*).

32. H.J. Ryser, *Science*, **150**, 501 (1965).

33. H.J. Ryser, *Biomembranes*, **2**, 197 (1971).

34. A.A. Yaroslavov, V.A. Kabanov, *Materials of All-Russian Kargin's symposium*, (2000) (*in Russian*).

35. M.M. Fel'dshtain, *Synthetic polymers of medical purpose. The 6th All-Union Symposium*, Alma-Ata (1983) (*in Russian*).

Conclusions

In this section we would like to state briefly the main results obtained by the authors with collaborators.

The novel guanidine containing monomer salts on the base of monomers of diallyl type:

- N,N-diallylguanidine acetate (DAGA),
- N,N-diallylguanidine trifluoroacetate (DAGTFA)

and of acrylic type:

- guanidine acrylate (GA),
- and methacrylate guanidine (MAG) were synthezised.

Their structure and configuration were established with the help of physical-chemical investigations. Their ability to radical homo-polymerization and copolymerization with diallyldimethylammonium chloride (DADMAC) in water mediums was studied.

We established that under radical polymerization of guanidine containing monomers of diallyl type – DAGA and DAGTFA in water and alcohol mediums the low-molecular polymers were formed that was caused by significant influence of degradational chain transfer to monomer on polymerization process. We showed that N,N-diallylguanidine acetate was able to radical copolymerization in water solutions with cationogenic monomer N,N-diallyl-N,N-dimethylammonium chloride (DADMAC) with formation of copolymers enriched by DADMAC.

As a result of systematic kinetic investigations of radical polymerization of acrylate and methacrylate guanidines in water solutions we found that in considered systems in the whole interval of monomers' concentrations the classic order regularities by concentrations of monomer and initiator were observed, the rates of GA and MAG polymerization and the values of intrinsic viscosities of resulted polymers were symbate and were increased with the rise of monomer concentration in the initial solution. Together with these facts we determined that homogeneity of reaction medium remained not in the whole interval of monomer concentrations but only in the following diapason of their concentrations: $0,.25 \le [GA] \le 1,30$ mole·l^{-1} and $0,20 \le [MAG] \le 0,4$ mole·l^{-1}. Molecular masses of some samples of mentioned polymers were determined by method of sediment ultracentrifugation.

Kinetic particularities of reaction of radical copolymerization of GA and MAG with DADMAC in water solutions were studied. We found that in both copolymerization systems rate of copolymerization was decreased with the rise of diallyl component content in the initial reaction solution. Resulted copolymers were enriched by acrylate comonomer in the whole interval of compositions of reaction mixture. Copolymerization constants for both systems and also intrinsic viscosities of synthesized copolymers were determined.

On the base of kinetic data obtained in the work we concluded that in reactions of homopolymerization the guanidine acrylate revealed high reactivity and in copolymerization the methacrylate guanidine was more active.

We showed that the reason of observed in systems with GA and MAG microheterogeneity was "denaturating" action of guanidine groups of monomer at their definite "excess" which were able to destruct the formed hydrated structure of macromolecules, formed during polymerization process that led to essential conformational changes accompanying in some cases by the loss of solubility and "salting-out" of polymer.

Bactericide and toxicological tests of synthesized in work novel guanidine containing homo- and copolymers on the number of cellular cultures were carried out. We demonstrated that copolymers revealed significant biocide activity in relation to E-coli and St. Aureus and maximum parameters were obtained in the case of copolymers with DAGA. Similar parameters were characteristic to copolymers with GA and MAG containing 30–70 mole % of acrylate component. We showed that homopolymer and a number of copolymers on the base of GA and MAG containing less than 40 mole % of DADMAC units were not toxic and at high content of DADMAC (up to 65–70%) were characterized by low toxicity. The exclusions were copolymers with DAGA which toxicity was close to DADMAC toxicity. On the base of data on biocidity we proposed variants of reception of non-toxic copolymers with high biocide ability.

Taking into account the perspectiveness of developing of synthesis of guanidine containing monomers and investigations of their properties including the ability to polymerization in laboratory headed by author the works in this direction are continued. We also use novel polymerization methods – various variants of pseudo-living polymerization which allows regulating fine parameters of resulted polymers (for example, structure, molecular mass, molecular mass distribution).

In conclusion we want to note the following. This book is prepared with the aim of systematization and presentation of the data obtained on radical (co)polymerization of novel guanidine containing monomers and comparison of these data with existing ones for monomers with similar structure to scientific community justice. Taking into account rapid development of scientific studies we may say that this is not final but intermediate result, determination of further perspectives and invitation to discussion. Author will take any advices, remarks and suggestions with gratitude. So, these are my coordinates:

Topchiev's Institute of Petrochemical Synthesis of Russian Academy of Sciences
Leninskii pr. 29, Moscow, 119991 Russian Federation
Fax: +7 (095) 230 2224
e-mail: sivov@ips.ac.ru

Appendix

In the Appendix we present some experimental details which may be useful for researches in practice.

Preparation of initial reagents and solvents

Allyl chlorous pure was dried and distilled over $CaCl_2$, $\mathbf{T_{boil}} = 45°C$, $\mathbf{n_d}^{20} = 1,4100$.
Diallylamine was distilled over NaOH in current of dry nitrogen before use, $\mathbf{T_{boil}} = 109,0–109,5°C$.
Tetrahydrofuran was preliminary stayed over the alkali, dehydrated over sodium in the presence of benzophenone. Tetrahydrofuran was used only freely distilled.
Acetone was dried over $CaCl_2$ and then boiled over P_2O_5 for two hours and twice distilled over P_2O_5, $\mathbf{T_{boil}} = 56°C$, $\mathbf{n_d}^{20} = 1,3550$.
Absolute ethanol. Ethanol (rectificate) was boiled for eight hours over fresh burnt calcium oxide (200g of CaO per 1l of ethanol). Then the mixture of 5g of magnesium shaving, 0,5g of sublimated iodine and 75ml of alcohol dried over CaO was distilled with backflow condenser until complete magnesium transformation into magnesium ethylate. Then the rest amount of alcohol treated by CaO was added into reaction vessel, mixture was boiled for 2 hours and alcohol was reduced.
Diethyl ether. It was dried over alkali and twice distilled over metal sodium. Stripped ether was kept over sodium, $\mathbf{T_{boil}} = 34–35°C$.
Chloroform was used as fresh distilled.
Acetonitrile was shaked up with CaH_2 and distilled over P_2O_5 in distillation flask with dephlegmator with high modulus factor of reflux, $\mathbf{T_{boil}} = 81,5°C$.
Trifluoroacetic acid was used as fresh distilled, $\mathbf{T_{boil}} = 72,4°C$.
Acrylic and methacrylic acids are technical products. For removal of water they were thrice freezed pouring out the solution thawed out at temperature lower than melting temperature of pure acid ($\mathbf{T_{boil}} = 13,5°C$ for AA and $\mathbf{T_{boil}} = 16,0°C$ for MAA). The residuary part of product was salted out over NaCl, dried over $CaCl_2$ and distilled over refined copper shaving under vacuum. For AA – $\mathbf{T_{boil}} = 26–28°C$ (3-5 millimeters of mercury), $\mathbf{d_{20}} = 1,051$, $\mathbf{n_d}^{20} = 1,4224$; for MAA – $\mathbf{T_{boil}} = 56–57°C$ (10 millimeters of mercury), $\mathbf{d_{20}} = 1,015$, $\mathbf{n_d}^{20} = 1,4315$.
Bi-distilled water with pH = 7 was used in all experiments.
Initiator purification. Pure ammonium persulfate (APS) $(NH_4)_2S_2O_8$, ammonium μ-peroxobis(tetraoxosulfate) was recrystallized out of bi-distilled water for several times and them dried in vacuum until constant weight.

Synthesis of initial monomers

1. Synthesis of diallyldimethylammoinum chloride

Synthesis of allyldimethyl amine (ADMA)

Necessary portion of 33% water solution of dimethyl amine (2,4 mole) is placed into two-liter four-throat vessel equipped by stirrer, backflow condenser and thermometer. After cooling of dimethyl amine down to 0–5°C the following reagents are added to it in the consequence: chlorous allyl (2,4 mole) and then 50% water solution of NaOH (3 mole). The rate of addition is regulated in such way to maintain the temperature not higher than 30°C. For rejection of heat excess the water bath with ice is used. After addition of alkali solution the reaction mass is slowly heated up to 60°C not permitting vigorous boil. It is stayed for 4 hours at 60°C. After cooling of reaction mass and removal of oily upper layer it is stayed over granulated NaOH for 40 hours. The fried amine is fractionated in column maintaining the modulus factor of reflux at 10 : 1. First of all the fraction of nonreacted allyl chloride T_{boil} = 45°C is gathered and then the fraction of ADMA with T_{boil} = 62-63°C. The yield of reaction product is 60%.

Synthesis of N,N-diallyl-N,N-dimethylammonium chloride (DADMAC)

Obtained at previous stage ADMA is mixed with equi-molar amount of dried and fresh distilled chlorous allyl in equal volume of dry acetone. The mixture is stayed for 2-3 days. Gradually at room temperature the DADMAC is crystallized in the form of white needle-shaped crystals. The obtained DADMAC is filtered, washed by dry acetone and dried in vacuum at temperature 40-50°C. then the reaction product undergo trice recrystallization out of mixture of absolute acetone and ethanol 90 : 10.

2. Synthesis of diallylguanidine acetate

In one-throat vessel (0,5l) 119,0g (1,22 moles) of diallyl amine is added and at cooling by ice and stirring 67ml (70g, 1,17 mole) of ice acetic acid is added. Then the solution of 49g (1,17mole) of cyanamide in 60ml of acetonitrile is added to obtained diallylamine acetate at stirring. Cooling is stopped and stirring is continued for 10 minutes. The resulted mixture is added by drops into three-throat vessel (volume 1l) with backflow condenser and dropping funnel to 250ml of boiling acetonitrile during 1 hour and 45 minutes and then it is boiled for one more 1 hour and 45 minutes. After addition of a half of mixture the DAGA crystals formation is observed. The resulted sediment is separated by filtration and washed several times with acetinitrile and acetone. It is dried in vacuum at 20°C. The yield of DAGA is 73%.

3. Synthesis of diallylguanidine trifluoroacetate

Synthesis of DAG

Metallic sodium (2,6g, 0,113 mole) is dissolved in 50ml of absolute alcohol and to the resulted solution of sodium ethylate at stirring 22,5g (0,113 mole) of solid acetate of DAG is added. Solution is stirred for 90 minutes and then stayed for 12 hours. After separation of sodium acetate sediment filtrate is steamed on rotary evaporator. The yield of DAG is 100% (15,7g), T_{melt} = 166–169°C.

Synthesis of DAGTFA

2,8ml (4,1g, 0,036mole) of TFAA is added by drops at cooling by ice to 5,0g (0,036 mole) of DAG in 20ml of ethanol, then it is stirred for 3 hours at 20°C and stayed for 12 hours. The solution is steamed on rotary evaporator. 8,2g of DAGTFA is obtained (the yield is 90%), T_{melt} = 156-157°C. for extraction and at additional purification one may carry out precipitation of monomer salt out of ethanol into dry acetone). If synthesis is carried out with trice excess of TFAA (7,4ml, 12,3g, 0,108 moles) 7,7g (yield 85%) of salt is obtained of the same equi-molar composition, T_{melt} = 156-158°C.

4. Synthesis of guanidine acrylate

Synthesis of sodium ethylate

Into three-throat vessel (volume 2l) equipped by stirrer and backflow condenser necessary amount of dehydrated ethyl alcohol (500ml) is added and at continued stirring metal sodium is added by portions during two hours. After addition of the whole sodium, reaction mass is stirred one more hour until complete dissolving of sodium.

Synthesis of guanidine

To obtained solution of sodium ethylate at constant stirring the equi-molar amount (1 mole) of guanidine hydrochloride (GHC) is added. Reaction mixture is stirred for 4 hours and stayed in refrigerator for the night. On the next day guanidine solution is filtrated from the formed sediment of sodium chloride.

Synthesis of guanidine acrylate

Reaction solution obtained the day before is placed in the vessel equipped by stirrer, thermometer, dropping funnel. After cooling of guanidine solution down to – 6-10°C at stirring the equi-molar amount of acrylic acid is dropped. The rate of addition is regulated in such way to maintain temperature not higher than 5-10°C. For rejection of heat excess cooling bath is used containing mixture of acetone and dry ice. After dropping of the whole amount of acrylic acid pH of reaction solution is close to 7. The solution is stirred for 4 hours at room temperature and then the vessel with reaction product is placed in refrigerator for several days. The obtained transparent solution of monomer salt in ethanol is precipitated into tenfold excess of dehydrated ether. Forming white needle-formed crystals of GA are filtered through compact glass filter, dried for several times by dehydrated ether and dried in vacuum at room temperature. The yield of GA is 80-85%. Then the synthesized salt undergo twice recrystallization out of mixture of dehydrated propyl alcohol and ethanol at ratio 90 : 10.

5. Synthesis of methacrylate guanidine

Synthesis of monomer salt on the base of methacrylic acid and guanidine was carried out by the technique described above for GA. The yield of MAG is 70-75%.

Structures and compositions of all synthesized monomer salts were confirmed by physical-chemical methods of analysis such as NMR and IR spectroscopies (see Sections 4.1, 4.2.1 and 4.2.2) and by elemental analysis (Chapter 1).

Kinetic measurements

Kinetics of GA and MAG monomers polymerization and also their copolymerization with DADMAC was studied by dilatometry method. Prepared reaction mixture was trice degassed on vacuum equipment (10^{-3} millimeters of mercury) in special ampoules after which the solution in vacuum was transferred into dilatometer and the system was blown by argon of special clearness. Then dilatometer was taken off the plant, closed by glass plug and placed into thermostat (**T** = 60°C). After thermostating the level of solution in capillary of dilatometer was leveled to necessary by removing of the part of it by syringe with long needle. The polymerization rate was determined by data on change of reaction solution volume with simultaneous registration of the change of solution level in dilatometer's capillary with the help of cathetometer KM-6 (the accuracy of measuring is ± 0,001 cm).

The degree of conversion of monomer into polymer was determined on the base of contraction values determined by densimetry method which for GA polymerization reaction in water was 10,8%, and for MAG – 7,0%.

The initial rates of polymerization were defined by tangent of inclination angle of initial linear sections of corresponding curves of dependence of transformation degree on polymerization duration. The relative error is not higher than 5%.

Extraction and purification of polymers from reaction solutions

Homopolymers and copolymers were extracted out of reaction mass by dialysis in relation to distilled water with the use of dialysis packets of "Spectrapor" firm (USA). Then free from monomers the solutions were replaced into crystallizing pans, water was evaporated (at room temperature), residual polymer was dried in vacuum drying oven over P_2O_5 at 40–50°C (for 40–50 hours), or out of water solution it was sedimented into acetone, then filtered and dried.

Synthesis of polymers and copolymers was carried out also by typical technique of radical polymerization in ampoules. Preliminary prepared solutions of monomers and initiator were replaced into ampoule which was trice degassed by freezing after which the ampoules were soldered and placed into thermostat for carrying out of homo- (co)polymerization. Then the water solution of (co)polymers were either treated by dialysis in water with the aim of removing of low-molecular products and then precipitated into cooled acetone, or they were precipitated into precipitator, dried and reprecipitated trice from methanol into ether or from water into acetone.

Viscosimetric measurements

Viscosimetric measurements of initial monomer solutions and model systems were carried out in viscosimeter of Ubellode type at 30°C. Intrinsic viscosities of polymers were determined in the same viscosimeter in 1n solution of NaCl in water at 30°C.

Physical-chemical methods of investigations

^{1}H NMR spectra were measured in spectrometer "Bruker MDS-300" (300MHz) in D_2O, Me_2CO-d_6, CD_3OD and DMSO-d_6 at 20°C, chemical shifts were determined in relation to residual protons of solvent.

IR-spectra of synthesized monomers and polymers were obtained on spectrophotometer "Spekord M82" (Carlzeis Jena). Samples for spectra measuring were prepared by pressing of tablets mixture of investigating sample with potassium bromide (1 : 200) thoroughly pulverized in agate pounder. The pressure of press was 4–6 t·cm^{-2}. The tablets obtained in such way were not cracked and remained transparent for several hours.

Thermo-physical method of investigation. Phase behavior of systems was studied on differential scanning calorimeter "Mettler" TA 4000 with heated cell DSC 30 at rate of heating 10degrees·min^{-1} in atmosphere of Ar.

Thermal stability was studied by method of differential thermal analysis (DTA) and differential thermal gravimetry (DTG) on derivatograph OD-102 (firm MOM, Hungary) on air at rate of heating 2,5 degrees·min^{-1} from 20 up to 500°C.

Elemental analysis was carried out by the method of pyrolysis chromatography on "Carlo Erba CHN-1108" (Italy).

Determination of molecular masses. Calculation of molecular masses (**M_{sd}**) of polymers was carried out on the base of data on high speed sedimentation and partial specific volume by Svedberg's equation:

$$\mathbf{M_{sd} = S_0 / D_0 \times RT / 1 - v\rho_0},$$

where: **S_0** – sedimentation constant, sec;
D_0 – diffusion coefficient, cm$_2$·sec^{-1};
R – gas constant ($8{,}32\times10^7$ erg·mole·$^{-1}$·degree^{-1});
T – temperature, K;
1 – **$v\rho_0$** – floatation factor;
v – partial specific volume, cm^3·g^{-1}.

The studied samples of homopolymers PMAG and PAG were preliminary purified from initial monomers by dialysis. Dialyzed water solutions of polymers were dried over P_2O_5 in desiccator, then in vacuum drying oven at $\mathbf{T} \leq 50°C$ until constant weight. For fresh prepared polymer solutions measurements were carried out in 1n solution of NaCl in water. As we established, low-molecular electrolyte of given concentration inhibited polyelectrolyte swelling of investigated homopolymers.

Sedimentation constants ($\mathbf{S_0}$) and diffusion coefficients ($\mathbf{D_0}$) were determined with the help of ultracentrifuge 3180 (firm MOM, Hungary) equipped by Filpot–Svenson optics allowing fixation the gradient of refractive exponent in time. Measurements were carried out at rotor's temperature 25°C. When selecting the rate of rotor rotation we based on preliminary determined values of [η] of studied samples (see the Table below).

Table. The values of $\mathbf{S_0}$, $\mathbf{D_0}$, $\mathbf{M_{sd}}$, $\mathbf{R_g}$ of studied samples.

No.	[M], mole/l	$S_0\times10^{-3}$	$D_0\times10^{7}$	$M_{sd}\times v$, $\times10^{-6}$	$R_g\cdot\pi m$	[η]
1	1 (MAG)	7,0	0,46	1,53	53,2	0,50
2	1 (GA)	4,9	3,0	0,17	8,2	0,27
3	2 (GA)	5,6	1,2	0,50	21,2	0,44

For PMAG sample the rate of rotor rotation (ω) was 38000 min^{-1}, and for PAG – 50000 min^{-1}. Determination of $\mathbf{D_0}$ was carried out at low rates (8000min^{-1}) in special cell.

$\mathbf{S_c}$ and $\mathbf{D_c}$ were determined at several concentrations for each of studied samples not exceeding values ~ 0,6 $g\cdot cm^{-3}$ at which the linear dependence of $\mathbf{S_c}$ and $\mathbf{D_c}$ on concentration of dissolved substance was fulfilled.

Diffusion coefficient ($\mathbf{D_c}$) for each concentration was calculated by formula (14):

$$\mathbf{D_c} = \Delta(\mathbf{Q} / \mathbf{H})^2 / 4\pi\Delta\mathbf{t} \qquad (14)$$

where **Q** – the area under the peak (cm^3);

H – the height of peak (cm);

t – the time of catrifugation (sec).

From the data of sedimental diagrams the $\mathbf{S_c}$ was calculated according to the formula (15):

$$\mathbf{S_c} = \Delta\ln\mathbf{X_{max}} / \Delta\mathbf{t}\omega^2 \qquad (15)$$

where $\mathbf{X_{max}}$ – the distance between maximum peak (sedimentation limit) from rotation axis of rotor (cm) at the moment of time **t**;

ω – angle rate (radian per sec^{-1}), $\omega = 2\pi\cdot\mathbf{n} / 60$; **n** – the number of turnes;

$\mathbf{S_c}$ – was determined for each concentration as tangent of angle of inclination of linear dependence lg**x** = **f(t)**.

Hydrodynamic radiuses of particles were calculated by formula (16):

$$\mathbf{R_g} = \mathbf{kT} / 6\pi\eta_0\mathbf{D_0} \qquad (16)$$

where η_0 – viscosity of the solvent (poise);

k – Boltzman's constant ($8,937\times10^{-3}$ poises).

Measuring of specific partial volume of the polymer **v** in solution was carried out by densimetry method. Determination was carried out in modified densimeter which neck was substituted by capillary at $\mathbf{T} = 25 \pm 0,05°C$ in water solution of 1n NaCl at polymer concentration in solution 1,0$g\cdot dl^{-1}$.

The values of **v** were calculated by formula (17):

$$\mathbf{v} = \mathbf{v_0}[1 / \mathbf{m_0} - 100 / \mathbf{p}(1 / \mathbf{m_0} - 1 / \mathbf{m})] \qquad (17),$$

where $\mathbf{v_0}$ – water number or densimeter volume, cm^3;

$\mathbf{m_0}$ and **m** – masses of solvent and solution in densimeter, g;

Density of solvent was calculated from the ratio:

$$\rho_0 = \mathbf{m_0} / \mathbf{v},\ g\cdot cm^{-3} \qquad (18)$$

The accuracy of determination of weight-average molecular masses $\mathbf{M_{sd}}$ from sedimental data is 5%.

The technique of estimation of bactericide activity of (co)polymers

For investigation of bactericide activity of synthesized (co)polymers the cultures of E-coli (colon bacillus) and St.aureus (goldish staphylococcus) from International collection of cultures were taken as test-microbes which microbe capacity was 0,1 ml of billion suspension per 1ml of preparation used at definite concentrations (twice dilution of preparation in sterile solution of distilled water). Preparations were not placed into control test-tubes. There were test-cultures E-coli, St.aureus and 1ml of distilled water in them. Exposition of preparations with test-microbes was 1 hour at room temperature. Then we made sowing from each experimental test-tube and from control one on nutrient agar Endo poured into sterile Petri dish. Then dishes with sowings were incubated in thermostat at 37°C for 18 hours and anti-bacterial activity of studied samples was determined.

The technique of estimation of toxicity of synthesized (co)polymers

This technique (in vitro method) determines the order of determination of toxicity with the use of short-time suspension culture of sells – bull's sperm. Estimation of toxicity is made under control of integrated products of disposable syringes and infusion single-use devices. Test-sperm is the granulated, frozen in liquid nitrogen vapors bull's sperm (frozen bull's sperm is obtained on the stations of artificial insemination). The principle of method is based on the analysis of dependence of mobility index of spermatozoa suspension on time:

$$\mathbf{m = f(t)}$$

Tthe index of mobility is determined as follows:

$$\mathbf{m = d \times C_n \times V},$$

where $\mathbf{d}$ – constant coefficient;

$\mathbf{C_n}$ – concentration of cells mobility;

$\mathbf{V}$ – the average modulus of cells rate.

Estimation of mobility index is made by calculation of number of fluctuations of intensity of scattered radiation caused by cells passage through optical probe.

$\mathbf{I_t}$ – index of toxicity which is determined by average values of mobility time ($\mathbf{t_{av}}$):

$$\mathbf{I_t} = (\mathbf{t_{av.exp.}} : \mathbf{t_{av.test}}) \times 100\%,$$

where $\mathbf{t_{av.exp}}$ – the average time of mobility of spermatozoa in test sample;

$\mathbf{t_{av.test}}$ – the average time of mobility of spermatozoa in control sample;

Values of $\mathbf{I_t}$ may be from 0 up to ∞ if the values of toxicity index of tested extract are in normal interval 60%÷120%, then extract is identified as non-toxic:

$$60\% < \mathbf{I_t} < 120\%$$

Investigation of polymers' fungicidal activity

For estimation of fungicidal activity of polymers against solid brand on winter wheat the plant was sprayed in the phase of the first unfolded leaf by solutions of testing samples (concentration was 0,075%). Controlled plants were not treated. Inoculation of plants was carried out by uredospores of solid brands on the other the day after sparge. Accounting of plants lesions by solid brand was carried out by calculation of a number of pustules on leafs.